Carlos Crovetto Lamarca was born in Concepción, Chile, in 1933. After finishing high school, he began his farming activities at Chequén Farm, in the county of Florida, province of Concepción. He received further training at the university level, and from the U.S. Department of Agriculture, Soil Conservation Service, which is now the Natural Resources Conservation Service.

Carlos Crovetto is widely known as a man who loves nature and, above all, the land he inherited. This land is his life endeavor and is characterized by his solid will of change.

Eager to learn and communicate, Carlos Crovetto has written more than 30 specialized articles on agricultural themes. These articles have been published in national and international magazines. He also has had many articles published in the newspaper *El Sur* of Concepción. In these articles he gives his opinion on regional matters concerning issues against the conservation of the essential resource, the soil. Because of his vast experience in no-tillage, he has been invited to a number of international meetings.

Since 1985, Carlos Crovetto has worked to promote no-till in Venezuela, Colombia, Argentina, Brazil, Uruguay, Bolivia, Mexico, United Kingdom, and the USA.

His activities are not restricted only to those already mentioned. In 1971, he became one of the founders and the first president of the Soil Conservation Society of Chile. He has served as its president since 1977.

Carlos Crovetto has been recognized for his work and influence in many areas. The honors include: the Honor Award and the Fellow Award from the Soil and Water Conservation Society, Ames, Iowa; the Agricultural Merit Award from the University of Concepción; the Applied Research Award from the city of Concepción; the Merit Award from the Monsanto Agricultural Company; and the Order of Agricultural Merit, the highest award given by the Ministry of Agriculture of Chile.

His work is summarized in *Stubble Over the Soil*. This book is the pinnacle of his desires to communicate and share his experience. At the same time it marks a new stage of change and openness in soil conservation.

STUBBLE OVER THE SOIL

The vital role of plant residue in soil management
to improve soil quality

STUBBLE OVER THE SOIL

The vital role of plant residue in soil management
to improve soil quality

Carlos Crovetto Lamarca

Photo A. No-till corn planted in wheat straw, 1994.
Photo B. Chopping wheat straw before planting no-till corn, 1994.
Photo C. Rainwater runs clear when planting with no-till system, 1993.
Photo D. No-till drill seeding oats on wheat straw, 1993.

1996

iii

Cover Design: Carlos Crovetto Lamarca

American Society of Agronomy, Inc.
677 South Segoe Road, Madison, WI 53711 USA

Reprinted in 1998.

Description of Cover Photo: No-till corn growing in wheat residue at
Chequén, 1989.

Collaborator and Scientific Advisor: Dr. Raúl Raggi M., Agronomist
Advisors: Dr. Hugo Zunino V.
           Gustavo Rojas L., Agronomist
           Juan Villa B., Agronomist
           Roberto Velasco, Agronomist

Library of Congress Catalog Card Number: 96-085124

Printed in the United States of America

To My Wife

Inseparable mate in so many
conservation events
I dedicate my deepest
feeling in this work.
With generosity and profound love,
she understood and encouraged
the writing of this book.

To My Daughters

May they forgive me as their father for not sharing with them
the time they deserved.
I beg them to keep the integrity of Chequén.
May this land, which witnessed their birth and growth,
mean neither hate nor division,
but the indivisible bond of all the family.

Copihue. Chilean National flower *Lapageria rosea* L.

The writing of this book is officially sponsored by the Ministry of Agriculture of the Republic of Chile.

Revised and Translated into English by:

Jerry Lemunyon, Fort Worth, Texas
David Schertz, Washington, DC
Lewis Daniel, Lakewood, Colorado
Stefanie Aschmann, Lincoln, Nebraska
Donald Baldwin, Enterprise, Oregon
Linda Oyer, Albuquerque, New Mexico
Maria Montes, San Juan, Puerto Rico

ACKNOWLEDGMENTS

I thank my parents and brothers for giving me the opportunity to work on Chequén since I was young.

My deepest thanks to the USDA Natural Resources Conservation Service for the teaching received during the 1950s and 1960s, which contributed to my formation as a conservationist. I also want to express my gratitude for their interest in this work, which undoubtedly has made this translation and publication possible.

To the translators and reviewers, my acknowledgment for their patient work, which will allow the image of Chequén and its productive conservation development to encourage farmers of other countries to a better understanding of our fundamental resource.

I am privileged to have very good friends in my life. Many of them have closely followed my work and experiences. When the idea of writing about my work and the relationship with the soil was born, experts in many fields, who were interested in my writing, volunteered their time with patience and dedication to this work. Their involvement in this project has enriched it.

The thousands of people who have visited Chequén in search of a better understanding of their own soils have been a big encouragement and have strongly influenced the materialization of this work.

My thanks also to the many foreigners who have contributed to my work with their encouragement; to the many Argentinean farm producers who have listened to me in extended conferences, especially those who have visited Chequén; the Brazilian and Uruguayan farm producers who, in response to my frequent travels to those countries, have arrived in large groups to observe firsthand the experiences told there.

CONTENTS

Chapter 2
The Management of No-Till

Chapter 3
Factors That Limit and Favor No-Till Production

Chapter 4
No-Till and Its Influence on Some Natural Parameters of the Soil

Chapter 5
Productivity and Profitability

PROLOGUE

I first had the pleasure of corresponding with Carlos Crovetto Lamarca about the theme of no-tillage 14 years ago. It was not until 1986, however, that I had the opportunity to meet him personally. That year, he brought a group of about 30 Chilean farmers to Kentucky to see our experiments at the University of Kentucky and the application of no-tillage by local farmers.

During his visit, I had the opportunity to see photographs showing examples of the work that he had done. It was impressive. At the time I had no idea of the difficulties he had working with the soils on his farm. Finally in 1988, I was able to go to his farm, Chequén, and see with my own eyes what he had accomplished in just a few years. The soils of the Cordillera on the coast of Chile are very similar to those in the Piedmont in southeastern USA. Both are formed in granite, and severe damage from erosion has occurred. Without a superhuman effort, it is practically impossible to return these soils to life. Nevertheless, Carlos Crovetto was able to renovate his soils and to obtain high yields of wheat and corn using no-tillage and adequate fertilizer. The soils have not only recuperated, but are far more productive than the original soils. At the same time, his farm has transformed into a beautiful landscape.

What are the principles of no-tillage? According to the author of this book, the forest is established in a natural way, without soil tillage, creating its own environment in the soil in order to grow. Perhaps, to a certain extent, no-tillage imitates the forest since the plow is replaced by the biological tillage of plant roots, which is gentle compared with breaking the soil with a plow. Furthermore, the residue protects the soil surface from the impact of raindrops. Between the roots and the surface residue, the structure of the soil begins to reform, just as in the case of a forest. With the use of fertilizer, the production of biomass increases and a new equilibrium is established. Little by little the nature of the soil changes, and with this change the water from rainstorms infiltrates the soil instead of running off. This results in even higher production of biomass.

Instead of a vicious cycle of soil deterioration, this cycle of increased biomass production, root activity, and protection of the soil is beneficial. More water gives more production, and more organic matter gives more water, which in turn gives more productivity.

In summary, a well managed system has the capacity to improve soils and the production of foodstuffs at the same time. This process is the theme of Carlos Crovettos' book. He has had success in the management of soils that were incredibly degraded. Perhaps more important, he has left his corner of the world in a much better condition than it was. In these pages readers can learn from his personal experiences. Each farmer and each professional has the possibility of following his example to improve the soil while producing more food. It is now fashionable to speak of sustainable agriculture. The truth is that many of the systems proposed leave the soil half worn out and produce lower yields than are desirable. The system of no-tillage, explained at great depth in these pages, presents an alternative that truly makes sense and which gives better yields and better soil at the same time. I recommend that you, the reader, study this work and consider how to apply the principles to other climates, other lands, and other countries.

Grant W. Thomas
University of Kentucky
Lexington, Kentucky

Dr. Grant W. Thomas in his first visit to Chequén Farm, 1988. A Contribution to the Soil and Water Conservation Revolution

FOREWORD

Stubble Over the Soil: The Vital Role of Plant Residue in Soil Management to Improve Soil Quality, as originally written by Carlos Crovetto Lamarca, who lives in Concepción, Chile, brings a different dimension to books published by our Societies. The subject matter is in the mainstream of the discipline, and the book covers a topic that is strongly based on the agronomic sciences, but it is written from the perspective of a user and its applications and perspectives originate from outside the USA and Canada. Furthermore, it is a translation, originally written in Spanish by a practitioner for an agricultural audience. Why then is it important to the Societies, and to its primary membership of scientists? The answer is in the comprehensive evaluation and application of technology in a way that proved to be successful. These documented successes deserve to be shared, along with the wealth of experience that was gained during the process.

We owe a great debt of gratitude to Jerry Lemunyon and David Schertz from the U.S. Department of Agriculture's Natural Resources Conservation Service (NRCS) who had the vision, drive, organizational savvy, and perseverance to first bring the book to the attention of the Societies and then to see it through to completion. Would the Societies be interested in publishing a translation of a comprehensive approach to successful adoption of a technology? The usual questions were raised regarding scientific value, credibility of authors, need for technical reviews, and expected sales. But the answers were different! The book was in Spanish, and already was written. What about scientific credibility? Could it be translated accurately and be technically sound? How would the Societies handle such a task? Will the interpretations be transferable? But times have changed and it was appropriate that our Societies accept the challenge and responsibility.

Principles of soil conservation and their applications have been discussed for centuries and the role of crop residues has been acknowledged. The volunteer system of soil conservation as a part of responsible stewardship of natural resources is legendary in American history. Now there is strong interest within the broader public, and the concept of soil quality is rapidly emerging. A new interpretation of this vital resource using new concepts is emerging. Sustainability and natural resource management are recognized as being critically important. It is time to take a new look and use a more holistic perspective.

Few books have been written in agronomy that are derived largely from on-farm trials and personal experiences. Following through with the usual thoroughness of our publication system, a draft was translated and reviewed. Then, the translation was checked and verified. Finally, it was edited capably by Matthew A. McMahon, Grant W. Thomas, Robert I. Papendick, Rattan Lal, and Gary A. Peterson (chair) and approved. Throughout, Carlos waited for the work to unfold, not being as interested in the recognition of his personal efforts as he was in the eventual use of the material. Now it is here, a new venture by our Societies, but a continuation of communication from the experienced to the less experienced, something that has been going on for centuries.

The topics covered in this book are of paramount importance. In no other time have people been as aware and as dependent on the need to conserve natural resources. Soil is one of those resources that is at the very foundation of our civilization. For that reason, perhaps more than any other, this book was written, not for the Societies, not even for the Northern Hemisphere. It was written out of love for the earth and the commitment to its preservation. All of us are learners and benefactors. We hope you enjoy the experience of learning from Carlos's experience in learning!

C. Jerry Nelson
President, American Society of Agronomy
A. Bruce Maunder
President, Crop Science Society of America
H.H. Cheng
President, Soil Science Society of America

PREFACE

To fully realize the potential impact on the reader of *Stubble Over the Soil* by Carlos Crovetto Lamarca, one first needs to understand some important points about the author. Crovetto is an experienced farmer. He first farmed with traditional methods, using the moldboard plow, until excessive soil erosion became such a serious problem that he realized the life of Chequén Farm was in jeopardy. In 1959, he eliminated the use of the moldboard plow and planted his entire farm to permanent grass and trees. He then studied in the USA, learning about soil conservation techniques, and developed a unique philosophy that should be embraced by all soil conservationists.

Nineteen years later, in 1978, Mr. Crovetto began implementing his new philosophy through no-tillage. What was this new philosophy? It was one of commitment. It all started with Crovettos' desire to improve soil quality, to make the best use of the land within its capabilities, and to achieve an economical return. Through his management techniques, the author achieved what we might say is the American Dream, planting row crops, forages, and trees; improving habitat for wildlife; and improving the quality of the soil while at the same time providing economical returns in an environmentally sound manner. What he has accomplished is a tribute to ecosystem management.

The reader will find that Mr. Crovetto is obtaining astounding yields on soils that would be expected to produce much less. In addition, he is protecting the soil from erosion, improving soil tilth and quality, and increasing the economic and environmental value of his farm.

In December 1993, I had the extreme pleasure of personally seeing the farming operations that have been implemented on Chequén Farm, near Concepción, Chile, which is owned and operated by Mr. Crovetto. This farm is truly a productive and sustainable farm. There is little or no soil erosion. In fact, because of the conservation system that has been implemented, approximately 1.0 millimeter of organic soil material is being developed annually as a result of leaving the stubble over the soil. He lives his philosophy that "the grain is for the farmer, and the residue is for the land."

This is truly one of the first books of its kind. The author combines technical information about the use of no-till or direct seeding techniques in Chile, the use of cover crops and tree planting, and includes an enlightening philosophy of the importance of soil conservation, soil chemistry, and soil quality. Many have written on this topic before, but primarily from a specific, single topic orientation rather than from a holistic viewpoint.

Mr. Crovetto has accepted, implemented, and expanded the philosophy of some of the greatest leaders of soil conservation, including Hugh Hammond Bennett and Walter C. Lowdermilk. He has certainly added significantly to their important contributions. He has addressed international groups and is a renowned speaker in Chile, Argentina, Uruguay, Bolivia, Brazil, Mexico, and the USA.

Before you begin reading this book, be prepared to learn from the technical material and philosophy that the author presents. It may change your thoughts, ideas, goals, or even your soil management techniques. But most importantly, it will make you stop and think about your personal soil and water conservation values.

David L. Schertz
National Agronomist
U.S. Department of Agriculture
Natural Resources Conservation Service

Introduction

*"The people who neglect their
land destroy themselves."*

ANA PRIMAVESI

This book is the negation of the plow and tools that cultivate the soil, the irrational burning of crop residue, and the deforestation of native vegetation on fragile soils. Much of the hunger and misery in the world is a result of the plow. Plowing causes erosion, which destroys the surface organic layer, resulting in lost fertility and capacity to absorb water. Actually, better means of producing food exist without using the plow and other tillage implements. Soil, a fundamental resource, also benefits when tillage is eliminated. My main motivation for this book was to share the knowledge gained during 40 years of working on my land, Chequén Farm.

The stimulus, as a result of visits to Chequén by farmers, students, professionals, and scientists, has made me delve deeply into this thrilling experience of writing a book. To this I should add the interest developed in other Latin American countries in no-till and the soil conservation we practice at Chequén. Writing this book was not an easy task, especially since my desire is to be outdoors in intimate contact with nature.

Since youth, I have felt the need to communicate. I do not like to travel alone because I cannot share my experiences. My first public speech was on soil conservation. It was presented at the Southern Agricultural Society in Concepción in 1961, a year after having made my first study trip to the USA. Because of my lack of experience in communication, I could not express my thoughts in an organized way. I was forced to read the talk. At this meeting I expressed for the first time the term no-till in vineyards.

Shortly after I wrote my first article in Concepción's newspaper, *El Sur*, I was motivated by constantly seeing the farmers returning home with bales of hay in their animal-drawn carts. It was easy to understand that their farm no longer produced sufficient forage to get through the winter, which necessitated them buying hay in the city. Erosion was destroying everything.

My work has led me down a different road. I had the opportunity, first, to observe my environment, and with it, the meaning of my presence on my land; and, second, to observe that what I did could mean a better future for me and for my fellow man.

To observe the production phenomena was of vital importance to establish comparatively the convenience of plowing or not plowing. Water erosion mortified me. I searched anxiously for a way to manage my land well, with the hope of making it better every day. To produce in that degraded environment was vitally important. This signified a tremendous challenge for I lacked sufficient knowledge to understand what was happening around me. I had a lot of fear at first. More than anything, I feared failure, however, I always had faith that things could change and that I only had to understand the process that had destroyed the soils of Chequén to know how to restore them. All of this was fortified by a strong confidence in what I was doing.

Together with a gradual increase in production, I soon observed that significant physical, chemical, and biological differences were developing in the soil, as evidenced by soil tests made regularly.

Group of Uruguayan farmers of FUCREA visiting Chequén (1989).

Field day at Chequén (1988).

Manoel Henrique Pereira and Franke Dijkstra, pioneers in no-till in Ponta Grossa, P.R., Brazil. The commemorative plaque on Franke Dijkstra's farm reads (translated to English):" 1976–1986. 10 years of no-till in the fields of P.R. On this farm, in 1976, no-till seeding was initiated in a green cover crop and with crop rotation. It represents a production system that is efficient in erosion control and increases productivity. Ponta Grossa, April 3, 1987" (1990).

Over the years, I have had the privilege to observe how erosion is controlled and how eroded soils recuperate when fire and cultivating equipment are eliminated. No-till was and is the effective alternative.

No-till is a new concept in the use and management of soils. It allows for sowing any seed without plowing or disturbing the soil surface. With no-till, traditional tillage implements, such as plows, disks, chisel plows, and various types of cultivators, are replaced by drills capable of cutting stubble and roots, leaving the seed properly placed in the soil.

Observing the processes of birth and death of animals and plants makes us aware that thousands of pastures and forests are an integrated natural work of mother nature. The bodies of plants and animals that have finished their cycle are involved in the processes of soil formation and help to generate fertile and productive soils. No-till largely approximates this ingenious work of nature since it in fact makes wise use of crop residue. The plow has no place in nature because it is replaced efficiently by biological activities. **Plants, roots, worms, arthropods, and other living beings are the best natural cultivators of the soil.**

No-till is the prolongation of the natural activities that produced forests, shrubs and pastures. On fragile soils, natural forests cannot be replaced by herbaceous or shrubby plants, no matter how vigorous they are; however, no-till is able to replace shrubs and pastures. The stubble and roots that remain on and in the soil after harvest are active participants in the genetic processes of the soil and soil improvement. Without a doubt, the greatest benefit of this system is that it makes use of crop residue to improve fertility and conservation of the soil.

The plow is, without any doubt, one of the oldest tools still in use. The same model of cast iron plow patented by Newbold in 1796 and perfected by John Deere in 1837 is still used and revered by farmers throughout the world. The moldboard plow was a relief from the effort that, until the 18th century, was needed to cultivate the soil by hand. This was a significant milestone in the history of man.

Since ancient times this tool has been directly responsible for feeding humanity. Its use is so tied to the idiosyncrasies of farmers that they still persevere in its use. For many farmers the plow is the pride and symbol of their work. This can explain, in some measure, why they continue to plow the soil. Another reason may be the fear of new things and of failure. The assurance that their traditional food will be available leaves the farmers without choices. Moreover, it is easy to understand that the deterioration of the soils in Chile and in the whole world has been caused by inadequate management of the agricultural tools used to work them. Management should certainly be more involved in the processes of conservation and improvement of the soils. **The countries and inhabitants of the Earth are suffering more so everyday, the demolishing effect of using the plow, in particular, and in the poor use of the soil, in general.**

The serious erosive processes on most of the cropped, pastured, and forested soils in the world began with the gradual loss of soil organic matter content. This loss was caused by deforestation, overgrazing, fallowing (plowing the soil several months before seeding), and the burning of crop residue. Consequently, the tools that have contributed relentlessly to the loss of the soil and its content of organic matter are **fire, the axe, and the plow.**

According to what is now known, the Earth is a privileged place in the universe because it combines all the conditions necessary to develop life; however, human activities gradually reduce the capacity of our planet to maintain life in a time when the increase in population and consumption creates increasing demands on that capacity. On the other hand, the conservation of renewable natural resources and the development of the world population have so rarely been combined that they frequently seem incompatible (even though sometimes one tends to say that they are compatible). They are incompatible as long as sustained development is generated without giving proportional attention to conservation.

In this respect, one can observe that the major damage to the soil in the world is centered in the most developed countries. People in these countries accept the term *soil loss* in daily life or, in the best of cases, prefer the term *sustainable agriculture*, **a definition that implicitly accepts soil loss as a typical function of agricultural production.** This incorrect agronomic standard does not acknowledge the contribution in fertility that natural agents are capable of producing through changes in the soil. In other words, the loss of fertility caused by the current or ancestral system used is not compensated by the natural capacity of the soil to self-generate. This equilibrium is not reached because the losses by anthropic origins are always greater than those that nature by itself can contribute. In most cases the effect of these soil losses are compensated by large applications of fertilizers and pesticides.

These negative processes are based on the lack of attention, priority, and defined policies that governments should give to universities, scientific centers, agricultural schools, students, and farmers.

The great efforts that science and research have made to provide farmers with improved technology for the use and management of soils cannot be ignored. Soil conservation, genetics, control of insects and diseases, fertilization, and irrigation have advanced considerably; however, all this technological development supporting conservation is not present on the land.

The classic problems generated by an intuitive agriculture give rise to serious disturbances in the management of the soils and cause erosion, sedimentation, compaction, salinization and acidification, ending with desertification of formerly fertile soils. This means that the centers of agricultural research could see their efforts blocked if they do not really transfer their technology for improved use and management of the soil. Proper technology transfer would

Typical landscape of the eroded areas of Florida County, Region VIII (1986).

The small farm has impoverished and destroyed the dryland soils of the country, keeping this sector submerged in poverty (Florida County, 1986).

require archiving valuable production advances, or in the best of cases, deferring their application at a high cost, until the techniques for improved use and management of the soil are implemented for the benefit of the nations agriculture.

In recent years, farmers in Chile have improved their production by using more efficient equipment and machinery, better weed control, more or better fertilizers and pesticides, and better plant genetics; however, processes with improved soil treatment may not necessarily be included. As yields increase, crop residue also increases. Therefore, an increase of yields should be synonymous with better soil conservation as long as farmers efficiently manage their crop residue and avoid the use of fire. Fires are common in Chilean fields and leave few possibilities of a more rational use of the soil. This constitutes an unexpected national catastrophe because of the enormous area involved, and it seriously affects its agroecological and productive aspects.

The National Forest Corporation has alarming figures on damage caused by agricultural and forest fires in the country. They indicate that annually an average of 60,000 hectares (150,000 acres) are burned, which has a direct cost in USA dollars of $20,000,000 and an indirect cost of $200,000,000 (Meza, 1991, personal communication). These figures do not include the land clearing by fire carried out in the scheduled burning program of CONAF.

In 1957 there were 18,870,000 hectares (47,175,000 acres) of land in the country that was eroded to varying degrees according to Decaraf, formerly with the Department of Conservation and Administration of Agricultural and Forest Resources. This constitutes 60.1% of the total agricultural area (Elizalde, 1970). These figures show only part of the problem, since the erosive processes have dramatically increased.

To increase production without having conservation as a base is relatively easy. The companies that commercialize agrichemicals are concerned with disseminating their products, so in many cases, they help the farmer to try them. This is certainly good as more can be produced at less cost.

The organization of farmers in Technology Transfer Groups, an initiative of the former Minister of Agriculture, Jorge Prado, has transferred the results of the research and knowledge of new technologies to farmers. This is achieved through the group organization taking into account the similarity of forms of operation (system of production) and the geographical location of the producers.

Presently, about 1800 farmers are organized in 100 Technology Transfer Groups. They manage a total area of 650,000 hectares (1,625,000 acres). The final objective of this group action is to produce an extension of the technology to the rest of the producers who do not participate in the system. It is estimated that each member farmer of a group extends directly to five producers and indirectly to 25 (Raggi, 1989).

Humanity, in search of economic development and the enjoyment of the natural resources, should face the reality of how limited the resources are, as well as the capacity and fragility of the ecosystems, and the needs of future generations for these resources. The purpose of development should be proportional to the social and economic well-being. The objective of conservation is, on the other hand, the maintenance of the capacity of the land to sustain that development and support life. A human civilization should base itself on these principles.

The two most important characteristics that distinguish our time are the almost unlimited capacity of man to build and create, and the opposite capacity of equal magnitude, to destroy and annihilate.

Solving the problems of soil management and conservation are not so easy now. A spontaneous change cannot be expected among the farmers who have maintained ancestral practices, and it is difficult to expect when there is no technical support to improve their knowledge of soil and

The plow is a typical implement of the small farm and one of the principal destructive agents of the dryland soils of Chile.

The hoe, which causes damage similar to the plow, is a much used implement on the small farm (Florida County, 1985).

water conservation. In the professional arena, the agronomists, agricultural technicians, and farmers are concerned with being successful in production and do little or nothing about conservation.

The phenomenon begins and develops in the universities. The students only study basic concepts of conservation, which is not sufficient. Chile is a country of mountainous relief, and it is there where we should practice our management of agriculture. Each worker of the land should have a better knowledge of the soil and the significance of its conservation. At the same time, the agronomists and technicians should lead in using better conservation practices that are appropriate with the topography of the fields, types of soils and crops and the amount of precipitation.

Chequén Farm gave me the opportunity to experiment with widely different agricultural production systems. For 34 years, the soils of Chequén have not been extensively tilled and for the last 17 years seeding has been done without any plowing or turning the soil surface. Many pioneer farmers in Chile and in the world are working their soil without plowing, but we need even more.

The need for food increases every day. The population increases incessantly, which requires a larger quantity of food. The natural properties of most soils under continued cultivation have been brutally altered. All of this, plus the fact that few virgin soils are available to bring into production, indicates to us that the perspectives for the future are not encouraging.

The population of the world is over 5.6 billion inhabitants, of which 4.4 billion live in the developing countries that have a 2.1% rate of annual growth. More than 1.2 billion live in developed countries, that have a 0.6% annual growth rate. With this perspective, by the year 2000 the estimated total population will be 6.2 billion inhabitants, with 4.9 billion living in developing countries with a growth rate of 2.0%. While more than 1.3 billion people are expected in the developed countries that have an annual growth rate of 0.5% (Ciancaglini, 1989). With this, one can appreciate that food needs in the near future will be greater and greater, which will require a more intense use of the soil.

Apparently soil conservation is at odds with food production because the worker of the land always has more immediate problems to solve than conserving the soil. The proper management of the soil, which also needs urgent attention, is left in second place.

I think that even the most simple farmer understands deteriorated soils and observes with fear how these soils affect production. We often hear: "The land is tired." That expression alone reveals the tragedy that millions of farmers experience throughout the world. The Earth is heavily damaged by the bad treatment it has been given for centuries. This situation can be interpreted as self-conformity. Each generation of farmers adapts unnecessarily to the degraded environment. This intellectual fusion of people with the environment applies towards the family, the community, and the country. For this reason, I will never tire of saying that **"the labor of a worker of the land is of concern to himself and his own community."** We farmers are only temporary workers of the fundamental resource and the well-being of our children. The future of humanity depends upon our conservation efficiency.

In the prologue to a pioneer work in Chile, *The Survival of Chile* by Rafael Elizalde, former Minister of Agriculture, Hugo Trivelli, wrote: "The conservation and restoration of the renewable resources of Chile is a task of all Chileans, without exception, without distinction of religion, creed, of partisan political convictions or economic interest. It is beyond all selfishness or sectarianism" (Elizalde, 1970).

This book by Elizalde signified a serious warning to governors and the governed since 1970 when the book was written. In spite of the time elapsed, a clear willingness to deal with the outcry is still not seen in the country. If it were not for the massive presence of Monterey pine, the present situation relative to conservation of natural resources, and especially of the soil, would be chaotic.

The combined destructive impact of poor human beings that fight for subsistence and the rich minorities that consume the major portion of the resources of the world is undermining the means that allow everyone to survive and flourish.

E.H. Faulkner, in my opinion one of the greatest conservationists of all time, warned in 1943 that the plow was carrying humanity to its own destruction (Faulkner, 1981). He generated a new conservation movement throughout the world that has obliged many specialists, agronomists, technicians, and farmers to more carefully observe and scrutinize the most used tool in agriculture—that from the simplest animal drawn plow to the most sophisticated versions of moldboard or disk plows.

Faulkner's work is still in effect mainly because what he maintained several decades ago is not a system of use and management of soils, but the conception of the significance of this resource. His philosophical proposal was based on natural order and equilibrium. He accepted at that time the use of the disk because agricultural implements that could seed into crop residue and modern herbicides were not available.

For several decades, the conservation of the soil and agricultural production on Chequén Farm have been based on the philosophy of Faulkner. One can say that no-till has proven him right, which magnifies his work. This system does without the plow, just like he said. This leaves many people perplexed and still others incredulous.

In our conservation work we have been touched by an enormous enthusiasm because we can see greater yields each year, as we continue to achieve what we so much yearn for—the complete conservation of the fundamental resource. One of the most important observations made in recent years shows the genesis of new organic soil rich in humus, black in color, and fertile. This phenomenon, not expected so rapidly, has changed the schemes, especially with respect to conservation, fertilization, and production.

Chequén sustains extraordinary increasing yields in wheat, corn, lupin, and oil seed rape (canola). These yields are a result of more efficient fertilization and better crop management. This would all be really promising if we could extrapolate, in greater measure, this agronomic situation to other places that are impacted by erosion and lack of productivity.

At the international level, Brazil shows a strong growth in its areas of no-till. In the 1970s, the increase was 400-fold, a spectacular figure (Derpsch, 1984). In fact, the locality of Ponta Grossa, in the state of Parana, is the major unit of no-till in the world. It covers an area of over 300,000 hectares (750,000 acres) and 90% of its area was seeded without plowing the land. This is an example that should be imitated by other countries. The following chart shows a clear tendency around the world towards an increase in areas seeded with the no-till system.

No-till continues to progressively increase in Chile and in the entire world. A notable example is being achieved in Argentina. A few years ago, no-till seeding was used on only a few hectares. Today, soil tillage is passing into history in such places as Rosario, San Jorge, Marcos

Evolution of no-till in the world (Derpsch, 1984).

Countries	Area cropped (hectares)	
	1973–1974	1983–1984
USA	2,200,000	4,800,000
Australia	100,000	400,000
Brazil	1,000	400,000
England	200,000	275,000
Japan/Malaysia/Sri Lanka	200,000	250,000
New Zealand	75,000	75,000
France	50,000	50,000
Holland	2,000	5,000

Juárez, Arequito, Venado Tuerto, and General Dehesa. In this respect, Dr. Victor Trucco, president of the Argentina Association of Direct Seeding, expresses that in the 1993–1994 season, 1,800,000 hectares (4,500,000 acres) were direct seeded in the country; only 3000 hectares (7500 acres) existed in 1987. This spectacular development of no-till in Argentina, especially in the humid plains (pampa), is because the soil is so strongly altered in its basic characteristics by water and wind erosion resulting from acute mechanization. The agricultural producers have no other alternatives. The truth is that minimum tillage or other forms of conservation tillage are not as capable as no-till in preventing soil crusting after seeding and an intense rainfall. These soils need precisely the opposite practices that have been used since the beginning of tillage. No-till, including leaving the stubble on the soil, is the alternative.

Many producers have understood this situation and presently use no-till as a source of conservation and permanent production. In this way they gradually recuperate the original fertility of the soil. Even though no-till has been used only a few years, producers can see a comparative increase in crop yields, which really stimulates them even more; however, the exceptional increase seen should go together with the research that tends to reduce all the negative effects caused by insects and diseases, weeds, and nutritional aspects of the soil and the plants, all of which certainly change when the soil is not tilled.

This technology is not easy to apply. It requires more knowledge, patient observation, and, more than anything, an irrevocable respect and love of the soil. Only in this way can the expected results be achieved.

In the rustic Andes Mountains, in the Magellan Steppes, on the slopes of the Andean Hills, in the humid plains (pampa), in the indomitable tropics, and in the fertile irrigated valleys, people of different origins, but united in one ideal, dedicate their heart and effort to their fellowman. This book is dedicated to all of these. I do this with the conviction that more than a message appears in these pages. It could mean a better understanding of nature and their work; thereby improving the form of life for the good of all humanity.

BIBLIOGRAPHY

Ciancaglini, N. 1989. *Análisis de algunas de las causas que producen salinización y problemas de drenaje en las árenas irrigadas. En:* Taller Técnico Examen de mecanismos de degradación y de metodologia en el manejo de aguas y suelos de tierras agrícolas bajo riego. Mendoza, Argentina. FAO/GCP/RLA/084/JPN.

Derpsch, R. 1984. *Histórico, Requisitos, Importancia e outras Considernaçoes sobre Plantio Direto no Brasil.* Cap. 1. Fundaçao, Cargill. pp. 2–3.

Elizalde, R. 1970. *La sobrevivencia de Chile. 2a ed. Ministerio de Agricultura.* El Escudo, Impresores-Editores Ltda., Santiago, Chile. pp. 20, 27, 95.

Faulkner, E. 1981. *La insensatez del agricultor.* El Ateneo. Buenos Aires, Argentina. 138 pp.

Raggi, R. 1989. *La agricultura de riego en Chile. Situación actual del riego, del drenaje y perspectivas.* En Taller Técnico, Examen de mecanismos de degradación y de metodologías en el manejo de aguas y suelos de tierras agrícolas bajo riego. Mendoza, Argentina. FAO/GCP/RLA/084 SPN. pp. 105–128.

CHAPTER 1

*Conserve
to
Produce*

After a harvest of pine carried out in 1962, this seedling of Monterey pine grows and develops spontaneously. This corresponds to the second generation established without human intervention, thereby forming a vegetative mulch 10 to 20 centimeters (4 to 8 inches) thick on the soil surface. Nature shows us the road; however, we are unable to understand its message (1991).

1.1. MY INHERITANCE

Like many people, I inherited a piece of soil on the face of the Earth, and like many of my colleagues, I have lived doing what is possible to nurture what nature provided.

At the death of my mother in 1953, the owners of Chequén Farm named me administrator of this 394-hectare (973-acre) property. The farm is in the middle of the Coastal Range at 280 meters (918 feet) above sea level and 35 kilometers (22 miles) east of the city of Concepción.

The property represents millions of hectares of dryland in the Coastal Range in South-central Chile. Chequén is 80% nonarable land in Capability Classes VI, VII, and VIII, which is suitable only for natural pasture and woodland. The remaining 20% consists of reservoirs, naturally protected areas, roads, and buildings.

Unstabilized gullies were products of the erosive processes started in the historical period of colonization of the country. In 1953, the gullies covered 11% of the total area of Chequén. These soils, converted by human activities into Capability Classes VII and VIII, were relegated to reforestation and wildlife without possibility of being incorporated into agricultural or livestock production.

Small areas in bottom lands where the surrounding watersheds converge correspond to Capability Class V. These soils had drainage and flooding problems in winter and were dry during the summer. They were eventually used for natural pastures and vegetable crops, depending on seasonal rainfall.

This description of Chequén Farm does not differ from that of neighboring properties or vast sectors of the eastern slope of the Coastal Range of Regions V to X.

Typical fallow tillage prior to planting wheat (*Triticum aestivum* L.) leaves the soil bare for a long time. This reduces the amount of organic matter in the soil, leaving it susceptible to erosion and loss of productivity (Yumbel County, Tomeco area, 1985).

Figure 1 is a map showing the original capability classification of the soils of Chequén Farm. These lands are the basis of my concerns and the laboratory on which I threw all my energy in search of the recuperation of the soil deteriorated by human actions. I did this to increase production in accordance with scientific and technically applicable principles. That was my challenge.

The map in Fig. 2 shows the historic use of the soils of Chequén Farm from 1953 to 1958 and the practices that are still applied on other farms. These practices caused severe deterioration of the soil, which resulted in very low production and the frustrated socioeconomic development of the regions. At the age of 20, I assumed the responsibility of making Chequén productive. My father had planted 50% of this land to Monterey pine (*Pinus radiata* D. Don), and had used the rest for natural pastures, wheat crops (moldboard plowed), and vineyards. A plow and disk pulled by oxen and a large hoe were used to cultivate the land. The wheat was threshed by horses on a threshing floor; therefore, the straw or stubble was removed from the place of harvest. This left the soil unprotected from the heavy winter rains.

I observed the ancient forms of production during my first 5 years of managing the farm. I realized that I should choose between studying the problem of soil management or quitting my job as farm manager.

A cooperative agreement was made between California (USA) and Chile during the 1950s. A large amount of resources was dedicated to soil conservation and agricultural production in Region VIII under a program called the Chillán Plan.

At that time I had contact with specialists in soil conservation. We started on Chequén with the seeding of dryland forages. The success reached with crimson clover (*Trifolium incarnatum*

Seedings done with traditional implements and without fertilization. Besides causing serious erosion, crop production is low (Florida County, 1986).

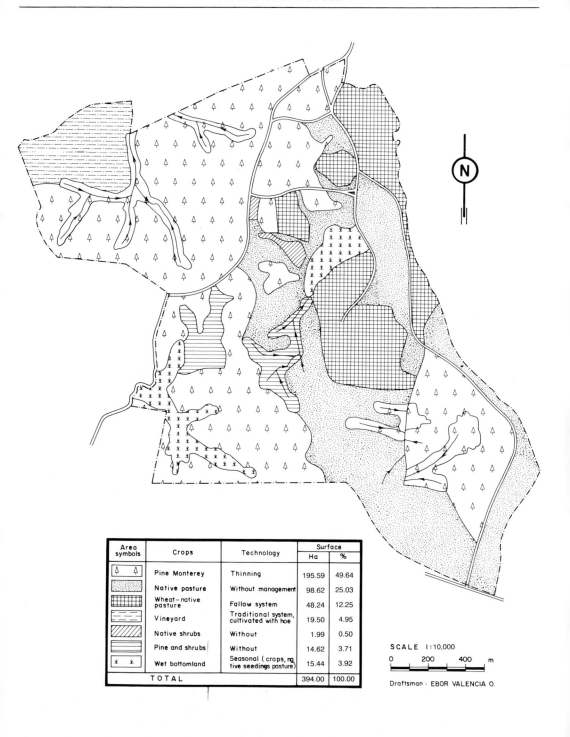

Area symbols	Crops	Technology	Surface	
			Ha	%
Δ Δ	Pine Monterey	Thinning	195.59	49.64
	Native pasture	Without management	98.62	25.03
	Wheat-native pasture	Fallow system	48.24	12.25
	Vineyard	Traditional system, cultivated with hoe	19.50	4.95
	Native shrubs	Without	1.99	0.50
	Pine and shrubs	Without	14.62	3.71
⚥ ⚥	Wet bottomland	Seasonal (crops, na tive seedings pasture)	15.44	3.92
TOTAL			394.00	100.00

SCALE 1:10,000

0   200   400   m

Draftsman : EBOR VALENCIA O.

Fig. 1. Land use on Chequén Farm, 1953 to 1958.

Seedings on land with steep slopes can result in serious soil loss by severe sheet and rill erosion (Yumbel County, Tomeco area, 1986).

Seedings on soils with moderate slopes also can result in serious damage by erosion (Mulchén County, 1990).

In dryland, the harvest of grain is done by hand, which requires that the straw be removed. This results in the loss of an important part of the organic material necessary for the soil (1986).

The sheaves are carried to the threshing area for threshing by horse or machine (1986).

Combined caption for both photos. Threshing by horse, the genuine celebration of the country folk, and by machine. In both cases, the straw is not returned to its place of origin, which further impoverishes the soil of organic matter, leaving it more susceptible to erosion (1986).

Soil tillage and the burning or removal of the stubble result in more runoff. This dramatically increases the water level in rivers, thus causing serious flooding damage (1960).

The first agricultural implement used on Chequén was an animal-drawn sickle bar hay mower. This was probably the first step towards managment of permanent pastures and no-till (1959).

A crimson clover pasture replaced traditional tilled wheat. The pasture was established for animal feed (1957).

Terraces on soils with steep slopes helped to reduce the erosive processes (San José Farm, Ránquil County, Courtesy of C. Grüebler, 1968).

The author observes the development of a seeded permanent pasture that replaced a wheat crop (1960).

Stripcropping was part of the first conservation system on Chequén. The results were unsatisfactory (1959).

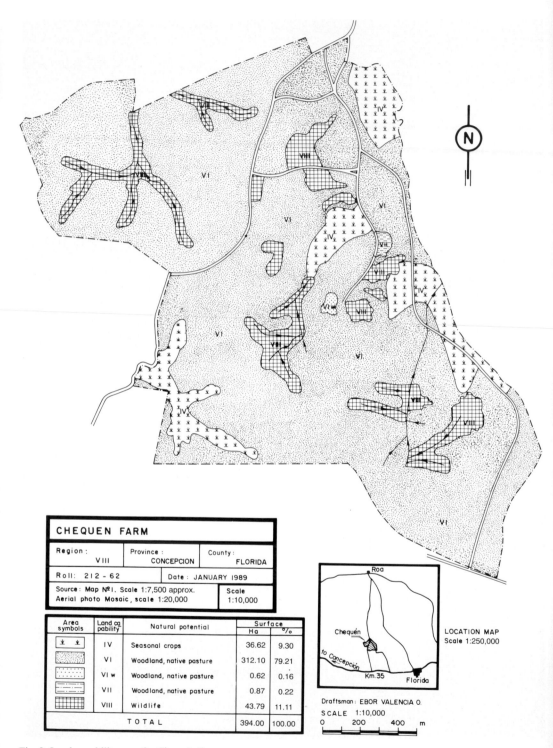

Fig. 2. Land capability map for Chequén Farm.

Wheat seeding on terraces on San José Farm, Ránquil County (Courtesy of C. Grüebler, 1968).

Vineyards in dryland under no-till (1959).

L.) as a producer of forage and the use of bitter blue lupin *(Lupinus angustifolius* L.) were vital to recuperate the fertility of the eroded soils and, and at the same time, increase their production.

Enthused by the initial success obtained, I received a partial scholarship in the USA from the Chillán Plan, which had training programs in cooperation with the Soil Conservation Service (now the Natural Resources Conservation Service) in different states. This began an important definitive stage in my life. I could apply new technology in soil and water conservation that was reinforced by the knowledge obtained in repeated trips abroad.

During the 1950s, there was great interest, especially by American farmers in the technology of Dr. H.H. Bennett. This technology included: construction of terraces, contour cultivation, stripcropping, and grassed waterways. The application of this technology allowed the farmers to reduce soil loss by water erosion as well as wind erosion.

On returning to Chequén Farm from my first study trip in the USA, I began stripcropping. I observed that the plowed strip did not have the same vigor and fertility achieved in the unplowed soil. The plowed strip was not exempt from erosion, its productivity was low, and the proliferation of weeds was excessive, which complicated even more the production of grain in those years.

All this indicated to me was that the technology of terraces and stripcropping was inadequate since erosion still persisted on these soils. This test led me to completely eradicate the use of the plow beginning in 1959, the year of my first trip to the USA. In August of that year, I stopped plowing and cultivating the vineyards. I was now disobeying my father's instructions, which caused me serious family problems. Instead, the weeds were cut with a sickle bar or superficially by a hoe.

To avoid the use of the plow in the seeded fields and continue with the traditional systems of soil management, I had to implement a system of supplements to the sharecroppers of the farm. Each year I provided them 80 kilograms of wheat for each child and another 80 kilograms for their wife. As a consequence of all this, I had to confront an additional cost that was absolutely unproductive in the short-term; however, this allowed me to have 100% of the farm in use for both conservation and production.

By replacing the plow and the growing of wheat, the improved pasture came about. In a short time this allowed us to manage an efficient number of Hereford cattle (for meat production).

In November 1978, 20 years later, I was able to establish the first no-till seeding of corn on Chequén, and probably the first in Chile. This happened after my trips to the USA and one to New Zealand in search of better agronomic knowledge.

I think it is important to point out that in the passing of those 20 years, I was the object of multiple pressures by my neighbors and even by my father who found my work absurd and irrational; however, I always had the support of my brother, Tomás, with whom I partially shared farm work since he had other responsibilities in the administration of Chequén. My feelings and acquired knowledge were solid. This allowed me to overcome the pressures, reinforcing what I wanted so much to achieve at a cost of considerable effort and sacrifice.

At that time I started a new life for me and my soil. Slowly the yields increased, and I observed with so much happiness how that soil mutilated by thoughtlessness, negligence, and irresponsibility, appeared as a valuable natural renewable resource—each day becoming more fertile and productive.

Figure 3 shows the current situation of the soils of Chequén and the improvements in different infrastructure necessary for the production from the soil. The fields of corn *(Zea mays* L.) are sprinkler irrigated, the water being elevated 45 meters (140 feet) from the reservoirs. The reservoirs are filled just by winter rainfall, which allows for irrigating 30 hectares (75 acres) of corn and 30 hectares of wheat stubble for winter forage or soil conditioning green manure for the next corn planting in rotation.

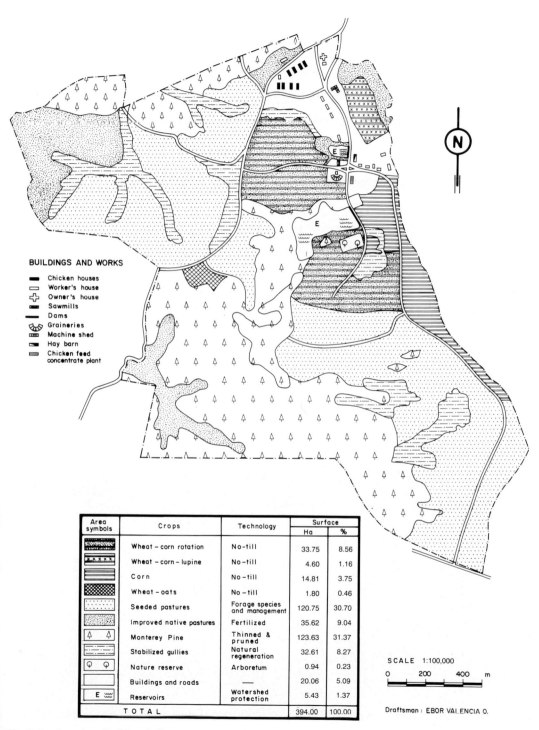

BUILDINGS AND WORKS

▬	Chicken houses
☐	Worker's house
⌖	Owner's house
▬	Sawmills
▬	Dams
⚙	Graineries
⊞	Machine shed
▬	Hay barn
▤	Chicken feed concentrate plant

Area symbols	Crops	Technology	Surface	
			Ha	%
	Wheat – corn rotation	No-till	33.75	8.56
	Wheat – corn – lupine	No-till	4.60	1.16
	Corn	No-till	14.81	3.75
	Wheat – oats	No – till	1.80	0.46
	Seeded pastures	Forage species and management	120.75	30.70
	Improved native pastures	Fertilized	35.62	9.04
	Monterey Pine	Thinned & pruned	123.63	31.37
	Stabilized gullies	Natural regeneration	32.61	8.27
	Nature reserve	Arboretum	0.94	0.23
	Buildings and roads	—	20.06	5.09
E	Reservoirs	Watershed protection	5.43	1.37
TOTAL			394.00	100.00

SCALE 1:100,000

0   200   400   m

Draftsman : EBOR VALENCIA O.

Fig. 3. Land use map for Chequén Farm.

Bench terraces constructed to use soils on steep slopes (1968).

In the 1960s and 1970s, permanently seeded pastures notably increased the number of cattle on Chequén (1968).

Table 1–1. Organic matter (OM), total carbon (C), total nitrogen (N), carbon/nitrogen ratio (C/N), soil reaction (pH), aluminum (Al), and extractable iron (Fe) in granitic soil profiles of Chequén Farm (Delgado, 1983).

Horizon	Depth	OM	C	N	C/N	pH	Al	Fe
	cm		%				parts per million	
		Native Forest (100+ Years)						
A	0–5	10.84	6.29	0.312	20.2	6.46	105	15.8
	5–10	9.83	5.70	0.265	21.5	6.24	194	15.7
	10–20	4.22	2.45	0.171	14.3	6.00	406	31.4
		Intensive Crops (50 ± 5 Years)						
ABp	0–5	2.28	1.32	0.092	14.3	6.02	127	35.6
	5–10	1.74	1.01	0.081	12.5	6.42	101	10.1
	10–20	1.52	0.88	0.072	12.2	6.32	165	10.2
		Forest (*Pinus radiata* (50 ± 5 Years)						
AB	0–5	3.22	1.87	0.067	27.9	5.89	230	36.6
	5–10	2.21	1.28	0.018	71.1	5.90	250	36.4
	10–20	2.17	1.26	0.020	63.0	5.77	325	36.5
		Permanent Pasture (improved 30 ± 5 Years)						
AW	0–5	9.48	5.50	0.290	19.0	6.49	73	10.4
	5–10	2.62	1.52	0.062	24.5	6.21	142	10.4
	10–20	1.16	0.67	0.052	12.9	6.19	131	47.0

In 1983, Luis Alberto Delgado, an agronomist, studied the effect of vegetative cover on the soils of Chequén under four different conditions for his degree thesis. The results obtained were of major importance in later establishing the comparative advantages offered by no-till with the passing of years. The analysis showed the values detected in the upper 20 centimeters (8 inches) of the soil horizons were the most dynamic and sensitive to changes.

The soil analyses shown in Table 1–1 were made in 1982, so the soils under no-till were not included.

The soils subjected to tillage show a large reduction in their levels of organic matter and, therefore, total carbon. The carbon/nitrogen ratio is narrower, which could indicate a comparatively lower biological activity. The soils with the highest pH show lower levels of extractable aluminum.

In the physical aspects, together with observing a radical change produced by water erosion, the particle-size analysis indicates large differences in the natural proportion of clay, silt, and sand.

The erosive processes activate the loss of the mineral constituents of the soil. The particles most affected are the finer ones because of their smaller size and weight. The percentage of clay in lowland native forest soils is 30.7%. Under intensive cultivation, it is 15.9%. On the other hand, the percentage of sand has increased from 46.8 to 66.2%, and the percentage of coarse sand analyzed increased from 2.78 to 41.91%, respectively (Delgado, 1983). These analytical results clearly show the degrading effect of erosion on particle-size distribution of the soil. It is especially shocking to observe the change produced in the sand fraction, where coarse sand increased dramatically in intensively cultivated lowland soils. Since coarse sands are inert material and comprise such a high percentage, it is obvious these soils were infertile.

This degrading effect of erosion is not just the heritage of Chequén. Millions of hectares of the Coastal Range are in this condition. Present generations must work on this degraded soil today. I think that no-till crops, pastures, and reforestation have a valuable role on our land.

Slowly, no-till approximates the ingenious work of the natural forest; but without pretending that it equals it. Our experience with no-till indicates that all the natural parameters of the soil improve compared with a soil under traditional cultivation. By improving the soil and its productivity, no-till comes out as the agronomic management that ensures the permanence in time of our fundamental soil resource.

1.2. WHY NO-TILL

1.2.1. Brief History of Soil Erosion

Human beings became farmers the moment they were able to sow and reproduce seeds, which happened about 7000 years ago (Lowdermilk, 1953). In that instant, one of the most relevant, historical deeds in the existence of living beings on Earth was noted. Together with ensuring survival, a sedentary life was initiated close to the land that produced the food.

The colonization of forested lands led to a gradual change in the use of the soils. These areas were transformed into pastures and small planted fields. Soon there were internal tribal struggles and wars between villages for better land, and damage to the conservation and productivity of the soil resulted.

Before Christ, in ancient Mesopotamia, great cities and civilizations used ingenious irrigation. Their fertile soils were eroded by the irrigation water and rain. In this century, former cities covered by sand and silt have been excavated, exposing the errors committed by past civilizations.

People are still not able to understand the lesson given by earlier civilizations. The more efficient the progress of the people, the more damage is inflicted on nature itself. It appears that people still do not learn from their own historic mistakes. An adequate knowledge of the management of renewable natural resources has still not been achieved (Fig. 4).

Deforestation, fire, and the implements that till the soil have been, without any doubt, the principle causes of a systematic destruction of the soil. The present generations appear to have their self-interest higher than the future existence of the human being. They do not realize the obvious risk that our descendants have in running out of reasonably fertile soil for food production.

Table 1–2 explains the severe ecological alteration that the soils subjected to traditional tillage have suffered compared with soils under no-tillage, where crop residue is left on the surface, and to soils under native forest.

This table clearly explains what happens in a soil covered with forest that has reached an equilibrium between its natural physical components and the organic mass. This was achieved by an exceptional equilibrium in a closed ecological system that was formed by inhabitants surrounding their environment. Once people used up the native forest, no-till should have been considered for use to repair the damaged soil. No-till is the ecological system that most approximates conservation perfection of the forest. This method of managing the soil includes a sustained ecological procedure capable of reactivating the biomass and absorbing the rains; thus reducing erosive runoff.

On Chequén, we were able to appreciate this phenomenon of repairing the soil while using the crop residue. The prolonged use of traditional tillage developed an open ecological system that was in severe disequilibrium. Characteristics of this system were a constant loss of organic carbon and nitrogen, minimal biomass, and inability to absorb rainfall; all of which provoked severe erosion. Figure 5 shows the losses and gains of nitrogen and carbon in relation to forms of management. These figures clearly show the serious losses of nitrogen and carbon that result from the destruction of the forest and the meadow by fire and the plow. These losses affect all the soil para-

In November 1978, the first seeding of no-till corn was established at Chequén, probably the first in Chile. Observing is Agronomist Roche D. Busch of the Soil Conservation Service of the USA (1978).

Erosion has degraded this soil to such an extreme that gravel larger than 4 millimeters (0.16 inch) in diameter appears on the surface.

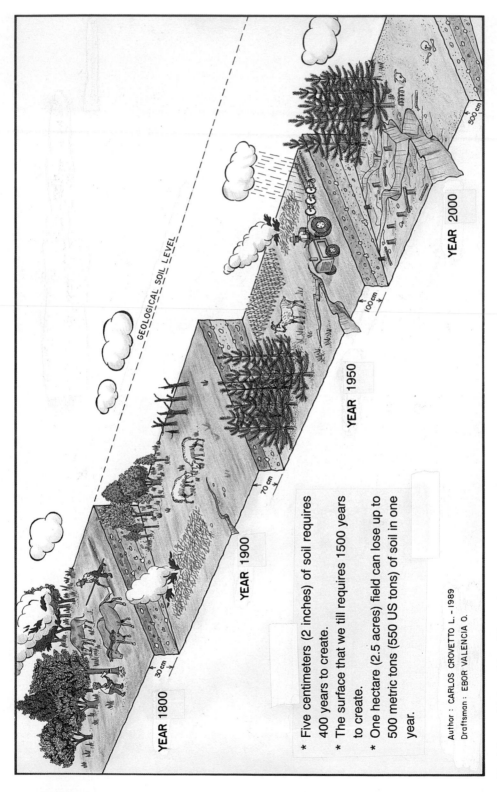

* Five centimeters (2 inches) of soil requires 400 years to create.
* The surface that we till requires 1500 years to create.
* One hectare (2.5 acres) field can lose up to 500 metric tons (550 US tons) of soil in one year.

Author : CARLOS CROVETTO L. - 1989
Draftsman : EBOR VALENCIA O.

Fig. 4. Progressive loss of the soil by erosion.

Table 1–2. Comparisons of ecological characteristics of a virgin soil and two tillage systems (Sierra, 1989).

Ecological systems		
Native forest	No-till with residue	Conventional tillage
Closed system	Semi-open system	Open system
In equilibrium	Sustainable	Unbalanced
High mineral recycling	Good mineral recycling	Low mineral recycling with losses of P, Ca, Mg, K
Maintains organic C and N	Maintains organic C and N	Losses of organic C and N
Biomass in equilibrium	Biomass actively growing	Biomass scarce
Optimum rainfall retention	Efficient rainfall retention	Poor rainfall retention
Minimal erosion	Slight erosion, strong soil forming activity	Severe wind and water erosion

meters. No-till with the use of crop residue can increase the levels of nitrogen and carbon to sustain life.

The uncontrolable increase of the human population and the lack of productive soils disturb many inhabitants. It is valid for us to ask ourselves: "Isn't there a technology applicable to agriculture capable of producing food without destroying our environment?"

1.2.2. Soil Conservation in America

In the preColumbian era, systems of soil protection based on terraces were practiced in the Andes Mountains. In Peru, it is estimated that of the one million hectares (2.5 million acres) of terraced lands, only 20% are still functioning. The rest are abandoned (Gischler and Fernández, 1984). Although it is true that these Andean terraces contributed to a better use of the soil, they

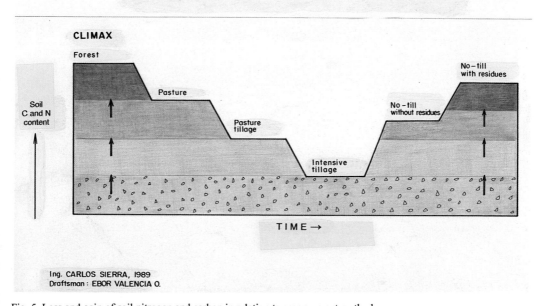

Ing. CARLOS SIERRA, 1989
Draftsman : EBOR VALENCIA O.

Fig. 5. Loss and gain of soil nitrogen and carbon in relation to management methods.

were only used by peaceful, sedentary, and well organized natives. The construction of these terraces required great ability and permanent care even though the conservation of the resource was not complete. In some measure, thanks to these management systems, these inhabitants made the soil a relatively durable property until our time.

Terraces are being promoted for protecting soils in Ecuador and other Andean Latin American countries. Without a doubt, this technology still has a place in our environment. Terracing provides a notable difference compared with systems that cultivate fragile soils without protective measures.

Lowdermilk also mentions terraces constructed in the south of France, probably by the Phoenicians about 2500 years ago. These lands, with slopes of 100%, still produce citrus using this conservation system (Lowdermilk, 1953).

In the 1930s, H.H. Bennett started a strong movement in the USA in soil conservation, which culminated with the formation of the Soil Conservation Service in 1935. At that time conservation systems that curbed erosion on millions of acres were initiated. After Bennett and Lowdermilk, F. Herbert, W. Austin and others developed new systems to maintain soil production. Without these systems the soils would have been converted to unusable gullies or, at the least, productivity of these soils would have decreased significantly. Observation of traditional conservation systems makes us think that they were designed only to reduce the erosion processes that these soils suffered when tilled. Therefore, we are only able to lengthen the fatal term of its destruction, and with it, the apocalypse of humanity.

It is estimated that worldwide between 5 and 7 million hectares (12.5 and 17.5 million acres) of soils are lost annually because of inadequate management. The projected loss of productive

Typical eroded landscape of the interior dryland of Region VII (Talca County, 1989).

Serious deforestation in the upper Duqueco River Watershed; Santa Bárbara County, 1983 (Courtesy of Dr. R. Raggi).

After reforestation with Monterey pine was completed, fire destroyed the vegetative mulch formed during development of the forest (Florida County, 1990).

Terraces are commonly used to manage the soil to partially avoid erosion; however, at the present time they are gradually being replaced by no-till.

Terraces must be well designed, constructed, and maintained because they tend to collapse (Río Grande del Sur State, Brazil, 1989).

Table 1–3. Soil loss under different management (Lombardi and Bertoni, 1975).

Soil management	Soil loss	
	tonnes per hectare	(tons per acre)
Soil plowed and seeded up and down the slope	up to 700	(312)
Soil plowed and seeded on the contour	400	(178)
Soil loss tolerance for a plowed, terraced soil seeded using conservation practices	12	(5)

soils by degradation is estimated at 100,000 square kilometers (38,600 square miles) by the end of the century (Marelli, 1989).

In the USA more than 12 tonnes per hectare (greater than 5 tons per acre) of soil is lost annually as a result of water and wind erosion on an area of almost 50 million hectares (125 million acres; Monsanto, 1988). In Alabama, soil loss by erosion exceeds 20 tonnes per hectare (nearly 9 tons per acre), which is equivalent to removing 1.8 mm (0.07 inches) of topsoil. According to the Universal Soil Loss Equation (USLE) of Wischmeier and Smith, it shouldn't exceed 5 tons per acre for that area (Yoo and Touchton, 1989).

In Paraná, Brazil, red latosol soils (Typic Haplorthox) that have a 6% slope show the relative efficiency of terraces in reducing soil losses; however, the losses are still too high for a sustained use (see Table 1–3; Derpsch and Alberini, 1982).

In Chile, Agronomist and Professor Luis Peña, established the erosion loss of "trumaos" soils of volcanic origin (Dystrandept). The soil tolerance is 7 tonnes per hectare per year (3 tons per acre per year), a loss that can vary according to the soil slope. This soil is in the Santa Bárbara Series. Trumaos is a local Indian name for dusty soils (see Table 1–4; Peña, 1986).

This is the present overwhelming reality of the soils of the world. Even worse, the degrading processes of the soil will continue to increase in the coming years.

The conservation techniques of Dr. Bennett, like all technology, can become obsolete. The contour terraces, together with strip crops, never offered a real protection to the soil; the farmers of that time simply accepted having lower soil losses.

Ernest E. Behn, an outstanding conservation farmer from Iowa (USA), states: "The majority of specialists in conservation believe that terraces are 50% effective in the control of erosion." He adds: "If a rain of 75 millimeters (3 inches) produces a soil loss of 18 tonnes per hectare (8 tons per acre), terraces on this same soil can reduce the loss to 9 tonnes per hectare (4 tons per acre). In exceptional cases, terraces can protect about 80% of the soil, which greatly reduces erosion; however, even using this technology, some fertile and productive soil will always run off toward the river and agriculture will be only sustainable in the short term" (Behn, 1977).

In climates where precipitation is extremely heavy, terraces used with no-till can reduce the transport of crop residue to lower areas.

The construction of terraces means a considerable movement of earth resulting in reduced productivity because the more fertile portions of the soil have been removed or buried. Agricultural machinery is obstructed by the terraces eliminating seeding operations in some areas.

Table 1–4. Soil loss tolerance of a Dystrandept according to studies using the Universal Soil Loss Equation (Peña, 1986).

Slope	Tolerance	
%	tonnes per hectare per year	(tons per acre per year)
4–7	8–10	(3.5–4.5)
8–12	7	(3.1)
13–18	5	(2.2)

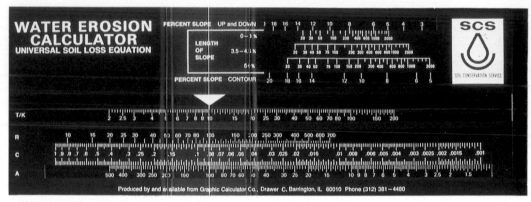

Universal Soil Loss Equation slide rule (1979).

Finally, construction and the maintenance required to prevent collapse of the terraces are expensive. The latter, however, is important.

1.2.3. Soil Loss Tolerance

Conservation must be clearly understood. The concept of soil conservation, as I understand it, is preserving the original characteristics of the resource. If this definition is not applied, the objective cannot be met.

What I intend to clarify is: "Is it tolerable or permissible to lose soil?" Unfortunately, the concept of soil loss, regardless of the amount, is so closely related to "conservation" that we end up thinking that **conservation permits the gradual loss of the basic resource**. At least this is what renown scientists have established by going to the extreme of designing a Universal Soil Loss Equation (USLE), which allows us to predict and quantify the probable erosion of a soil that has known physical characteristics and particular management practices.

This way of thinking, so commonly known and used in soil conservation, puts great pressure on farmers and students of agronomy who, using this concept, **learn to live with erosion,** without ever considering the possibility of being able to produce without destroying.

Many experts in soil conservation maintain, arbitrarily, that a soil loss can be up to 12 tonnes per hectare (5 tons per acre) per year, without greatly altering the productivity of the soil. I think that science and technology have gone too far with a whimsical criterion that goes beyond our reality. The entire world cannot afford to lose any more soil than it already has. Continued soil loss cannot be justified whatsoever because for the past two decades new management systems have been available that allow permanent use while maintaining and even improving the fundamental characteristics of the soil.

I firmly believe that the concept of soil loss, mathematically expressed by the USLE, developed as a necessity because of a deficiency in the traditional soil conservation systems. Specifically, it estimates the loss from sheet and rill erosion from a cultivated soil under various conditions using conservation practices, such as terraces, contour farming, or strip cropping. Then, the specialists, preoccupied by the potential losses, designed a system that would at least allow them to reduce the severity of the soil being lost. Undoubtedly, this has allowed them to improve conservation systems, but not to escape some intolerable loss, however small it may be.

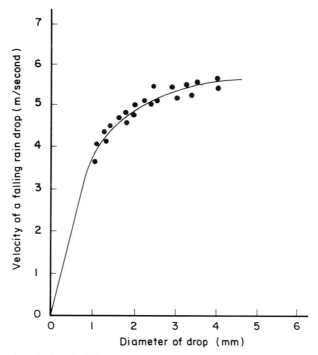

Fig. 6. Relation between the velocity of a falling raindrop and its diameter (Marelli, 1983).

According to the progress made in the use and management of soil, and especially in the change of criteria that has existed since the work of E.H. Faulkner, we can no longer tolerate the idea of losing soil. Furthermore, and contrary to what USLE predicts, it has already been demonstrated that new soil can be generated if agronomic systems are used that do not alter the soil structure. This is true even with high productivity. **This should indicate the need to develop a predictive tool that calculates how much soil organic material is regenerated instead of how much soil is lost** (oral presentation by the author at the IV International Conference on Soil Conservation in Maracay, Venezuela, 1985).

1.2.4. The Raindrop

Observing the impact of a raindrop on bare soil helps to understand many phenomena that have exhausted the soil over thousands of years. A raindrop is several times larger than a soil particle, whether mineral or organic. Considering its size and the gravitational energy it carries, a raindrop becomes a projectile that violently impacts bare soil exposed by the mechanical action of tillage implements (Contreras, 1975). The most important processes of anthropic degradation of nature are initiated by splash erosion caused by the water drop.

The kinetic energy that a raindrop develops in free fall can be used to measure its erosivity. This factor depends on the diameter of the drop. Beals, as mentioned by Marelli (1989), indicates that a drop 1 millimeter (0.04 inches) in diameter can arrive at the soil with a velocity of 3.5 meters per second (11.5 feet per second), while a drop of 5 millimeters (0.2 inches) can arrive at 5.5 meters per second (18.2 feet per second; Fig. 6, Marelli, 1983).

The erosive processes are initiated when the raindrop hits the soil surface. Soil aggregates consist of the union of primary soil particles (sand, silt, and clay), united by organic and mineral colloids (organic matter and clay). Where the aggregates are impacted by the raindrop they separate into smaller fragments and even into individual primary particles. The fragments are carried by excess water that flows over the surface, causing sheet erosion. These individual particles, especially those that correspond to the silt and clay sizes, plug the soil pores when they settle, thereby reducing the natural rainfall infiltration capacity. Runoff and, therefore, loss of fertile soil result.

1.2.5. The Advance of Gullies

Gullies constitute the final expression of erosion or destruction of soil and life. They are deep channels that are from a few meters in width, depth, and length to so large that they cover extensive areas. Their peripheral growth causes irreparable losses of agricultural soils each year if measures are not taken to stabilize their evolution. Part of the lost soil is deposited in low fertile areas, which leaves them unproductive. Gullies can form in just a few decades depending on the frequency with which the fields are plowed, type of soil, slope, existing vegetative cover, and the quantity and intensity of rainfall.

Dr. Wilfried Endlicher established that the number of gullies in the district of Tomeco increased from 420 in 1943 to 550 in 1973. Tomeco is 35 kilometers (22 miles) from the city of Concepción and 25 kilometers (16 miles) south of Chequén (Endlicher, 1988).

The total area of gullies that affect agricultural soils in Chile is estimated to be similar to that in the area of the Province of Concepción, which is 568,110 hectares (1,420,275 acres). With the exception of the area of Tomeco, Yumbel County, Region VIII, a high percentage of this area is reforested. This indicates that the soils in this area are at least partially stabilized.

It is interesting to listen to the renowned Professor Hernán Contreras M. expound on the problem of erosion, especially in Region VIII. He stated:

"The majority of wars, of the conflicts that originate between nations, have as a cause the defense of their territory. When a foreign threat looms over a country, all of its inhabitants arise as one person to defend the native soil. Constant struggles are produced by border situations where many times one fights for a few meters of worthless rock. Millions of men have given their life for this cause."

"However, a silent enemy, but cruel and constant, threatens us day and night, taking like a stealthy thief our wealth, the patrimony that we are so willing to defend, leaving unerasable footprints by its passing. In being visible to us, it would appear that it wouldn't change us, as if the land, which now is not there, wasn't ours, the sustenance of today and the bread of our children."

"Half of our productive soil is gone. It was taken, our sovereignty is threatened, our dependence is greater each day, obligating us to beg for our sustenance from other nations of the world that still have resources. The Bío Bío River has taken, in the last 100 years, the soil equal to 5 billion cubic meters (176.5 billion cubic feet), which loaded on 10 ton trucks, bumper to bumper, would form a line that would go 12 times around the world" (Contreras, 1973).

The plantations of Monterey pine have diminished peoples fear and the psychological pressure that comes from seeing the gullies. A bare soil is depressing. In this way, Monterey pine has transformed the landscape into an exceptional means for soil conservation, and at the same time, transforms desolate areas into places with a living and productive nature; however, one should consider that the presence of a gully, although forested, shows the permanent absence of a formerly agriculturally productive soil. The gully will continue in its frightful advance each time that the trees are harvested and/or the residue burned.

Rainfall simulator constructed at Chequén shows the erosive processes in soils under traditional tillage, pasture, and no-till and the sedimentation in Containers 1, 2 and 3, respectively. The detail of Tray 1 (insert) shows the effect of the impact of raindrops on soil under traditional tillage (1991).

Top: Soil impacted by raindrops shows severe surface crusting that impedes plant emergence and the absorption of rain-fall. Bottom: Excessive tillage of this soil and the lack of organic matter on the surface, together with intense rainfall, result in serious reduction of germinating seeds and low yields (Rosario, Argentina, 1989).

This soil, under traditional tillage, shows the crusting or flattening of the surface by the effect of raindrops (Bahía Blanca, Argentina, 1991).

After an intense rain, the germination and population density of this stand of corn are seriously affected by the lack of residue on the soil and by excessive tillage (1967).

Soils under the no-till system do not crust after a rain. These corn plants show good vigor due to the benefits of the system (1986).

In Chile, reforestation with Monterey pine has protected hundreds of thousands of hectares against erosion. Coelemu County, 1984 (Courtesy of Dr. R. Raggi).

Volcanic soils are characterized by their gentle hills and extensive areas suited to grain, livestock production, and woodland (Mulchén County, 1989).

Sedimentation of the Lonquén River at the outlet into the Itata River brought about by centuries of water erosion in the area, Trehuaco County (1990).

We must first understand that the soil lost by erosion, in its different expressions, is that which deposits in low areas and reaches creeks, small streams, rivers, and finally the sea. In a large measure, wind erosion that occurs on the coast and forms actively moving dunes, is the product of the sand transported by the rivers that the sea returns, covering useful coastal soils. It also is important to add the contamination of water by solids, such as clay, silt, fine sand, organic matter, nutrients, and agricultural pesticides. This problem is especially serious because of sedimentation occurring in rivers and bays in most of the country. The water that has high indices of solids in suspension causes severe disturbances in potable water treatment plants. Their operation is made more difficult by the presence of particles finer than 0.002 millimeters.

1.2.6. Complete Conservation

I believe that no-till acquires a new inspiration or philosophical concept that is based on the complete conservation of the resource. It is destined precisely to create a new organic soil over the degraded one. This is only possible where crop residues accumulate on the surface and are managed with minimal incorporation or soil movement. No-till is incompatible with intermittent tillage and any attempt to turn the soil to incorporate excess stubble, control insects, etc., leaves the soil in the same condition it was in before no-till.

The management of agricultural crop residue has been made into a real art on which the success of no-till largely depends. How fast a new organic soil layer forms depends on the type and amount of residue, type of soil, agronomic management, and climate. The wider the carbon/nitrogen ratio, the longer the humic state of the organic material will be sustained. No-till not only permits conservation, but also contributes to form new soil and increase sustainability of its productivity.

On Chequén, 35 years after abandoning the plow and tillage tools, the use of the soil is very different. The land use map before the application of conservation norms and no-till corresponding to the period 1953 to 1958 (Fig. 1) and the map of the recent use of the soil in 1988 (Fig. 3) show important changes. Among the most important change is native pasture improvement through management techniques and the establishment of dryland forage species, such as hardinggrass (*Phalaris tuberosa* var. stenoptera (Hack. A.S. Hitch.), ryegrass (*Lolium* sp.), and subterranean clover (*Trifolium subterraneum* L.).

The gullies, which were so typical of the landscape of Chequén, have stabilized. Their advance was impeded by the management of native reseeding and/or plantation or natural reseeding of Monterey pine. Bulldozers were used to fill large gullies in a total area of 10 hectares. Sprinklers now irrigate these soils. The incorporation of these soils into production had a high initial cost that was amortized over 5 years. Today, the yields are similar to those of neighboring soils; however, the cost of production is slightly higher because more nutrients must be added. The reclamation of these soils not only increased the area suitable for planting, but also improved the access for the different agricultural machinery in use. This aspect contributes notably to a greater efficiency and economy in the management of the crops while reducing the hours of work per hectare.

Of the land dedicated to traditional crops of wheat and oats (*Avena. sativa* L.) in rotation with natural pasture, 14% has been incorporated into the no-till system in rotations of wheat–corn, lupin–corn, wheat–lupin, wheat–canola (*Brassica napus* L.), and corn monoculture with a green cover crop [oat and vetch (*Vicia* sp.)] during winter. The natural and improved pastures make up 40% of the area of the farm. According to the results obtained, the area of no-till planting is constantly increasing, demonstrating the efficiency of the technologies applied.

Table 1–5. No-till yields of corn–wheat rotation in relation to rainfall, fertilization, plant population, and residue.

Year of Seeding	Crop	Precipitation	Nitrogen	Phosphorus	Yield	Residue	Plants
		millimeters (4)	kilogram per hectare		tons per hectare		per ha
1978	corn (1)	1350	160	92	4.6	3.8	55,650
1979	wheat	1080	90	46	2.2	1.9	—
1980	corn (1)	1303	220	92	6.5	6.0	59,232
1981	wheat	1164	100	46	2.8	2.2	—
1982	corn (1)	1633	260	92	7.8	2.2	62,305
1983	wheat	866	120	00 (3)	4.3	3.5	—
1984	corn (1)	1243	360	46	11.6	12.4	87,650
1985	wheat	865	140	00 (3)	6.2	5.8	—
1986	corn (1)	1433	340	46	8.5	9.1	88,125
1987 (2)	wheat	1059	140	00 (3)	4.8	4.1	—
1988 (2)	corn (1)	1019	315	46	12.2	14.7	82,321
1989 (2)	wheat	850	130	46	4.4	3.9	—
1990	corn (1)	799	305	58 (5)	12.5	13.8	81,204
1991	wheat	1083	270	46	7.5	7.6	—
1992	corn (1)	1391	405	66	13.3	14.6	96,500
1993	wheat	1046	200	50	4.8	5.6	—
1994	corn (1)	917	418	60	13.7	15.2	93,300

(1) Sprinkler irrigation from October to February.
(2) Urea replaced by sodium nitrate.
(3) Phosphorus not applied because the soil had 51 parts per million.
(4) Annual precipitation of Chequén.
(5) Ammonium phosphate and sodium nitrate replaced by phosphate rock and calcium magnesium ammonium nitrate, respectively.

On Chequén, the soils under no-till have shown an exceptional response to the conservation management that the system involves. New soil organic material forms at the rate of 1 millimeter per year, improving the physical, chemical, and biological condition of the soil. This is a consequence of the continuous addition of organic material through the stubble and of not tilling the soil.

Figure 7 shows the effect of no-till on the soils of Chequén. The information in this figure illustrates the reason I consider the no-till system as the most appropriate and effective conservation technique to improve the soil known to date.

To improve the range of economically feasible alternative crops and make use of the watershed on Chequén, two reservoirs were built to store winter rainfall. The reservoirs total 350,000 cubic meters (12,355,000 cubic feet). This has made spring seedings of such crops as corn, sunflower (*Helianthus annus* L.), and lupin possible, and has resulted in a gradual increase in production. The yields are shown in Table 1–5.

After 14 years of no-till on Chequén, it has been shown that even dry years can be profitable for seeding wheat or other grains. No-till can store more rainfall than soils under tillage. This is especially important in years with a dry spring, when the mulch on the soil can make the difference. During September, October, and November of 1989, the rainfall on Chequén was only 70 millimeters (2.8 inches). The average rainfall in this period is generally 217 millimeters (8.7 inches). Even with precipitation a third of normal, it was possible to produce 4.4 tonnes per hectare (66 bushels per acre) of wheat. From this, one deduces that if it is true that the quantity of rain during vegetative development of plants is important, other factors, such as erosion, weed control, fertilization, soil pH, and pest and disease control, can more seriously affect the established crop.

The yields obtained on Chequén are supported by agronomic management that is improved each year. The exceptional yields are the result of perfecting our knowledge of weed, insect, and

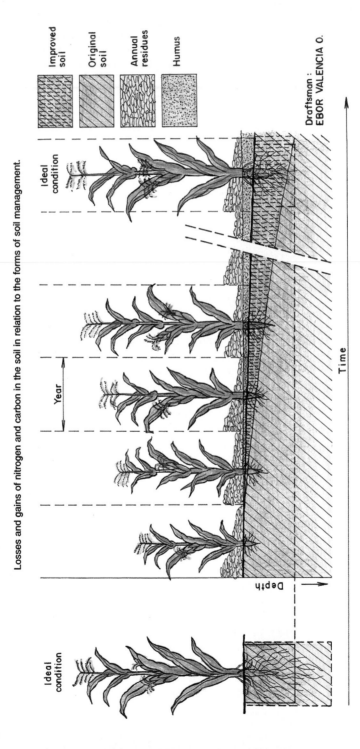

Fig. 7. Idealized scheme of the effect of no-till on the improvement of the physical and chemical properties of mineral soil in relation to time.

Farm tractors were used to fill in gullies on Chequén (1990).

Heavy machinery is often used to fill in gullies (1982).

Construction of a diversion channel at the head of this gully will keep water from running directly into the gully and increasing its size (1991).

The first Chequén reservoir has a capacity of 300,000 cubic meters, which allows for irrigating 30 hectares (74 acres) of corn (1978).

disease control, and fertilization, as well as paying more attention to the phenomenon of vegetative development of the plants. Also, it is important to point out that without the use of no-till these results would not have been achieved. Without a doubt the gradual improvement of the yields resulted from the regenerated physical, chemical, and biological conditions of the soil promoted by no-till.

In the following chapters, the successes and failures in the achievement of no-till management are discussed. The job has not been easy because sufficient experience supporting this special activity does not readily exist at the universities or agricultural research centers. Therefore, all the experience that I intend to provide is limited by the lack of external support. Perhaps, the major motivation that I have had to write this book has been precisely the fervent desire to provide specific knowledge in an environment that is lacking it. I am completely convinced that the management of no-till searches an idealized and humanized agriculture in which the center of the productive game is the farmer; **a laborer, hard working man, an untiring worker who now can manage the soil without the hateful dust or mud that for centuries made their hopes perish.** Now they can be dignified by their work.

1.3. SUITABLE SOILS

In general terms, all soils are adapted to no-till; however, it is appropriate to clarify that the ideal soils are those that facilitate the seeding, root, and vegetative development of the crops.

No-till seeding on a fine textured soil that has low content of organic matter would not initially present favorable conditions for the development of the plants. Clay textured soils are very sensitive to the lack or excess of moisture, either of which can harm the crop. Medium to coarse textured soils facilitate the seeding and permit a good germination.

The main objective in using implements that cultivate the soil is to achieve a good seedbed. This objective is only valid during a short time. The physical–mechanical turning done by tillage tools is achieved in an absolutely artificial manner, and as such, its duration is minimal. When the rain falls on bare soil, it causes not only the erosive processes already mentioned, but also compaction. The realignment of the particles form a hard crust on the surface of the soil. Tillage also breaks the natural channels made by plant roots, requiring that the soil be repeatedly cultivated. Soils with excess moisture, fine texture, and low organic matter on the surface aggravate the problems of root development. Because of this, most farmers insist on using the plow and tillage tools.

Perhaps for this reason no-till is at a comparative disadvantage with conventional tillage, especially in the first years the system is established. This is especially true, as has already been stated, in soils that have fine texture and low levels of organic matter.

When farmers initiate no-till, they should look for the best soils. The wetter and finer textured soils are generally in the valley bottoms. Planting in these soils is not recommended without having more experience in the management of no-till. On the other hand, planting on sloping soils, even though they are more eroded and have a lower level of organic matter, has shown good results, even better than the wet bottom land.

The lack of moisture in heavy soils, such as the clayey Alfisols of the Coastal Range, can be critical in the management of the system; however, better root development is achieved with adequate fertilization, especially nitrogen, at the time of seeding or within a few days of emergence. A crop in a more fertile medium can achieve greater root depth and, therefore, obtain more water and nutrients.

In fine-textured soils, organic matter content can be more important than the type of clay and its proportion in the soil. Therefore, no-till on a heavy soil can give good results if its organic matter content is above 4%. Generally, this condition is easy to find on nondegraded pastures and where excessive trampling by livestock has not altered the microrelief of the land. At least 1 tonne per hectare (900 pounds per acre) of mulch or organic residue on the soil surface benefits the first-time no-till seedings.

Using no-till on a soil that was previously under tillage reduces porosity for a time and, therefore, the oxygen content as a consequence of the compaction already mentioned. This situation is common in bottomland soils that are wet and have low organic matter content and is particularly difficult for the establishment of no-till. Considering these characteristics, it would be ideal to precondition these soils for future planting by improving the drainage and stimulating the existing vegetation by fertilizer application.

The soils in Chile most suitable for no-till are those of volcanic origin, or trumaos. These soils represent an area of about 4 million hectares (10 million acres), of which about 800,000 hectares (2,000,000 acres) are classified *arable*. The crops that can be established in these soils vary because of the geographical distribution, the wide range of climates, and the variety of soil capability classes. Because of this, these soils have a great agricultural importance (Espinoza, 1973).

The cultivation of wheat covers about 375,000 hectares (937,500 acres) of trumao soils. These soils make up almost 50% of the national wheat-growing area. This soil is used to produce 80% of the oats grown in Chile and also is used for sugar beets (*Beta vulgaris* L.) and canola.

No-till in Chile has increased since 1982 to an area estimated at 100,000 hectares (250,000 acres). Even though most farmers in southern Chile are enthusiastic practitioners of conservation, no-till has experienced slow growth because the stubble is burned. Without stubble on the soil, the no-till system does not function properly. And, ironically, burning of the stubble does not lower their production costs.

The Trumao soils of the Andean foothills have exceptional physical qualities, great water infiltration capacity, and excellent moisture retention. These soils have high organic matter content and allow planting at velocities exceeding 8 kilometers per hour (5 miles per hour). There is a rapid and uniform emergence of crop seedlings all of which should be positive in the efficient management of no-till. This phenomenon is discussed further in Chapter 2.

1.4. SUITABLE CROPS

When no-till is initiated, the general optimum conditions to establish a crop are not present, but this improves with time. Some crops have better adaptation than others and produce better results.

The adaptation of crops to no-till is influenced by pests and diseases. These problems, which can damage the crop, can be stimulated during the first crop years by the presence of residue. Good management of crop residue, however, can prevent damage to the established crop. This can be one of the sensitive points in no-till management. A regular and timely observation of the crops by specialists can help overcome the initial damage and even eradicate the problem. The biological control of pests and diseases in no-till crops is very encouraging. The crop residue, paramount to this system, also can stimulate beneficial organisms in such control.

In many countries where no-till is common, it is basically developed with a corn–soybean rotation. This rotation is a success in the USA's Corn Belt, Brazil, Argentina and other countries where it is done. If this rotation is performed longer than 6 years, pests and diseases can be produced, complicating the management in such a way that the rotation gradually starts to deteriorate.

The corn–soybean rotation is not common in Chile because soybeans produce a lower profit than corn and other crops. A description of the most important crops with no-till follows.

1.4.1. Summer Crops

1.4.1.1. Corn (Zea mays L.)

Corn is one of the greatest crops that can be adapted to no-till. In fact, in addition to soybeans, it is the most widely no-tilled crop world-wide.

In Chequén, corn was the first no-till crop. It was planted in October (spring) and harvested in April (autumn). Corn was first planted in a meadow and in the following years in the preceding corn–wheat rotation residue.

The corn plant is very sensitive to weed competition. It grows very quickly where there are no weeds and when the crop is well fertilized. At Chequén, the average measured growth of this plant is 4 centimeters (1.5 inches) per day. Between the last week of December and the first weeks of January, growth can be up to 10 centimeters (4 inches) per day, depending on the temperature in that period. Consequently, this plant can reach a height of 240 centimeters (94 inches) in 60 days.

Corn generally tolerates the monoculture very well. Significant changes do not occur in the comparative yields of the following years, although differences have been registered in production costs.

In 1984 corn monoculture, planted with winter oats as a green covercrop, produced 21,200 kilograms per hectare (337 bushels per acre) in a half hectare plot. This remarkable production indicates that the Pioneer 3747 (MR) corn in this latitude is very well adapted to no-till and has great genetic potential if a higher plant density is intensely managed. These yields probably do not coincide with those expected from the Pioneer variety, which is considered semi-early and has a moderate productivity.

I attribute this high yield to the nutrients and organic matter accumulation in the soil surface. Clipping the green vegetative cover 45 days before corn planting provided nutrients and organic matter to the soil. Another factor that could have influenced the high yields is the high population achieved, which was 106,000 plants per hectare (42,900 plants per acre). The soil and its physical management were adequate and the general agroclimatic conditions were favorable for the high yield of corn.

The increase of carbon dioxide in the lower atmosphere that surrounds the plants can stimulate the plants development. The atmosphere generally contains 0.03% of carbon dioxide. Because corn is a C4 plant, any small increase is significant. This suggests that the increase in carbon dioxide generated by the respiratory process of the soil microbiology stimulated by the high quantities of surface residue can improve the corn plants development and increase the yields when synthesis of more carbon is achieved in photosynthesis (Ciba Geigy Agrochemicals, 1979).

Corn planting is of great importance in the pedogenetic process. Its exceptional root system penetrates the more compacted soil layers and facilitates slow infiltration and retention of rainwater. Herbicides are available that leave the soil clean from weeds, helping the following crop in rotation. Finally, the great quantity of stubble produced by the grain crop leaves good vegetative cover for the following winter. At Chequén, corn is the crop that best adapts to no-till. The yields obtained through the years are shown in Table 1–5.

1.4.1.2. Soybean [Glycine max (L.) Merr.]

It is peculiar that the cultivation of soybeans is not important in Chile. It seems that the lack of interest is more economical than agronomic, since some varieties are well adapted to the south-

The plants with severe chlorosis result from the soil being saturated with water for a long time (1989).

No-till corn planted for silage production (1991).

Sprinkler irrigation used on no-till corn planted in crop residue (1986).

Corn plants are well adapted to the no-till system. Under good management, corn plants produce high quantities of adventitious or aerial roots that help obtain better nutrition (1990).

Oats and vetch planted in the fall for green vegetative cover are cut and chopped 45 days before the corn is planted. This rotation increases soil fertility, weed and pest control, and soil moisture conditions.

Pioneer 3747 corn variety, no-till planted in the same area as shown in the photo above, produced a yield of 21.2 tonnes per hectare (337 bushels per acre; 1983–1984).

An increase of carbon dioxide in the atmosphere of the corn canopy produced by a higher microbiological activity in the no-till soil may be responsible for higher yields.

A successful no-till soybean crop after wheat stubble on the farm of the well-known Argentine farm producer José Capretto (Colón, Argentina, 1989).

central zone of the country. This valuable legume is widely used in poultry and hog feed, and has to be imported as soybean meal. Like most legumes, its need for nitrogenous fertilizers is limited, which reduces costs appreciably. Soybeans also are a great rotational crop.

The results of no-till soybean production at Chequén have not been totally satisfactory compared with most of the other legume crops; however, this does not mean that growing this crop should be avoided. The main problem with soybeans and other legume production generally comes from the soil surface residue. This residue is an ideal environment for slug (*Agrolimax reticulatus*) proliferation. Slugs seriously damage soybean crops in the first growing stages (see Sections 3.4.2.1.). The bean fly larvae (*Hilemia platura*; see Section 3.4.2.2.), also attack both the seed and seedling. The lack of soil mechanical cultivation also aids the development of the slugs and larvae.

Soybeans should be used in the summer crop rotation with no-till because of the net benefits for the next crop.

1.4.1.3. Sunflower (Helianthus annuus L.)

The sunflower oil crop has shown to have a good adaptability to no-till. It is a vigorous plant with good root system growth when the soil has adequate moisture. Because sunflower is a broadleaf plant, broadleaf weeds are difficult to control. Therefore, the crop should be planted in soils free from broadleaf weeds using systemic preplant herbicides. Control of grassy weeds, annual or perennials, is very simple using nonselective postemergence grass herbicides.

Sunflowers are well adapted to the no-till system and is an excellent crop for rotation (1989).

Wheat planted in 12 tonnes of corn stubble per hectare (10,716 pounds per acre).

No-till wheat in oats stubble, Montpellier Farm (Mulchen County). Shown from left are the farmer, Roberto Parragué, and
  his son Rodrigo, Ko de Reujter, Dutch agronomy student, and Pedro del Canto, consultant agronomist (Mulchén County,
  1990).

Wheat growing in corn stubble.

The wheat in corn stubble is close to boot stage and spike emergence.

No-till planted wheat on the Mario Meyer Farm, which is close to Osorno, Region X. Meyer is an outstanding crop producer. Inspecting the crop is Roberto Aichele, consultant agronomist and pioneer in the introduction of the no-till system (1989).

1.4.2. Winter Crops

1.4.2.1. Wheat (Triticum aestivum L.)

After 16 years of no-till planting, wheat has become the most important winter crop. This crop has good adaptability, especially when it is planted in lupin residue or corn stubble. In the summer of 1994, Novo INIA wheat planted with a no-till system yielded up to 10.7 tonnes per hectare (159 bushels per acre) with an average on a four acre field of 8.2 tonnes per hectare (122 bushels per acre). This spring variety was drilled late in winter and was sprinkler irrigated on soils with 30% slope.

Wheat in no-till is sensitive to the lack of nitrogen. This particular stand received 750 kilograms per hectare (670 pounds per acre) of calcium magnesium ammonium nitrate (27% nitrogen) during the seeding and growing, plus 600 kilograms per hectare (536 pounds per acre) of North Carolina phosphate rock (30% P_2O_5) before drilling.

Other minor cereals, such as oats and triticale, are more adaptable to no-till planting than wheat; however, other management aspects must be considered. They include rotation, residue management, planting time, species and quantity of seeds, nitrogen fertilization, and weed, disease, and slug control.

1.4.2.2. Oat (Avena sativa L.)

Agronomically this grain has great adaptation qualities to different soils and climates. It has resistance to most pests and diseases, especially to the ones that attack the root.

The oat–wheat rotation is now a tradition in Chile, where the farmers can obtain a good production in the stubble of both crops. Because oats have abundant foliage development, this grain is recognized as a plant that keeps weeds under control. Because of these agronomic characteristics no-till planting of oats is very important for a good rotation. Its germination under these conditions is fast and vigorous, responding better than other crops. Another common observation made by the farmers is the capability of oats to absorb the phosphorus that is unavailable to other plants. This remarkable characteristic may be a result of its capacity to exude enzymes at the root zone, obtaining phosphorus that other plants normally cannot assimilate.

At Chequén, oats were introduced in areas under irrigation. It is an obligated winter crop in the no-till corn monoculture. Planting has been simplified to such a point that only a fertilizer spreader is used to broadcast the seed. The seed nestles between the corn stalks, and is later covered with the residue left by the chopper.

The capacity of oats to germinate without being buried is remarkable. A little straw over the seed and soil is enough for good germination. The chaff protects it, and it germinates only when moisture and temperature conditions are adequate. Fertilization is done with the same spreader used to sow the seed and is preferable early in the fall or at the end of winter to avoid leaching losses by nitrogen. The green stubble provides better weed control and soil nutrients, benefiting the summer crops. This way a new corn crop begins, avoiding in some ways the problems of monoculture.

In excessively wet soil during the winter, the cereal grain–legume rotation does not exclude the use of oats as a green cover during that season. The vegetative control of oats is carried out using systemic herbicides 20 days before planting the corn or other crops. Contact herbicide has not shown good results controlling oats and weeds typical of the spring season.

Combining oats no-tilled in wheat stubble (1989).

Cutting an oats–vetch crop to improve soil fertility and land preparation for no-till corn (1990).

Roberto Parragué, crop producer and pioneer in the introduction of no-till on volcanic soils (trumaos), observes the vetch–oats crop cut for animal feed at Chequén (1990).

1.4.2.3. Lupin

1.4.2.3.1. White and Yellow Lupin

The diversity of sweet lupins has been studied by Erik von Baer, a geneticist from Region IX (Temuco). Among the more cultivated varieties are Moltolupa, Llaima, Astra, Giant, and *L. luteus* L. var. aurea Baer. These varieties planted with a no-till system adapt well in clayey soils; however, they respond much better in lighter soils such as *trumaos* of the foothills of the Andes Mountains. The climatic conditions of Regions VI to X are adequate for its development although low precipitation can mean lower yields. In crops north of Region IX, weed control is more important in years of low precipitation. Temperatures below zero degrees Celsius (32°F) can seriously affect these legumes. Figure 8 shows the lupin developmental cycle and its relation with climatic characteristics (Baer, 1986).

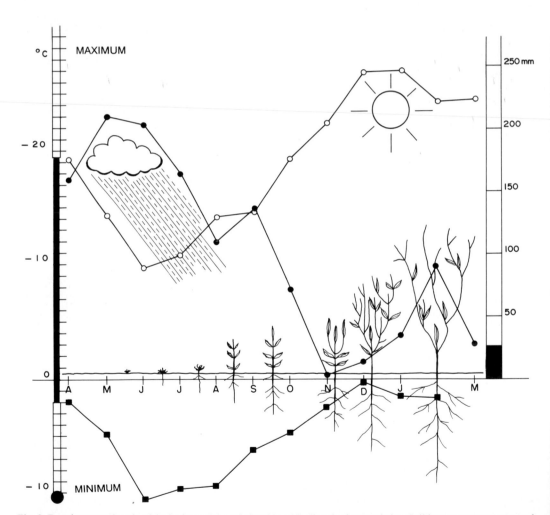

Fig. 8. Developmental cycle of the lupins and the relationship with climatic characteristics. Solid squares represent monthly average minimum temperature in degrees Celsius. Open circles represent monthly average maximum temperatures in degrees Celsius. Solid circles represent precipitation in millimeters (Baer, 1986).

Sweet lupin planted in corn stubble (1990).

During mechanical harvesting, we see substantial losses by pod shelling because dehiscent pods are formed during the blooming stage. The pod, when matured, twists, causing its seeds to shatter, an action that is stimulated during the threshing activities. A higher threshing speed can lower the shelling losses (Baer, 1986).

At Chequén, both the white lupin (*L. albus* L.) and the yellow lupin (*L. luteus* L.) had moderate adaptation over corn and wheat stubble because of its poor initial establishment. This plant, like other legumes, is attacked by slugs (*Agrolimax reticulatus*; see Section 3.4.2.1.). For this reason more seeds (160 kilograms per hectare; 143 pounds per acre) are used when planting. Disinfecting the seeds with Lorsban (Dow), Marshall 25 TS (Hoechst), or something similar, helps to control the larva that could attack this legume.

In a no-till system, the *L. luteus* var. *aurea* has been shown to be useful in forage production on degraded meadows. In the locality of Cajón, Region IX, a fall seeding at the rate of 80 kilograms per hectare (71 pounds per acre) can produce three forage grazings before the crop blooms in the spring. This significantly contributes to the cattles winter feed (Baer, 1989 personal communication).

1.4.2.3.2. Narrowleaf Lupin (*Lupinus angustifolius* L.)

This variety of Australian origin (Geraldton, Western Australia) is for winter and spring crops. It was recently introduced to the country and shows good cultural characteristics. Its white color and small seeds, similar to the soybean, have a good adaptation to no-till in the clayey soils of Chequén. Excellent germination was observed, using seeds disinfected for fungus and soil insects. The plants were vigorous from the beginning, showing a strong and active nodulation. The seeds were not inoculated because the soil was previously treated with the bitter blue lupin (*L. angustifolius* L.). The vigorous nodulation seen in the plant roots reflect an excellent nitrogen con-

Table 1–6. Chemical analysis (% dry matter) of the sweet lupin (*Lupinus angustifolius*) seed.

Tests	Rough Seed	Shelled	Hull
Moisture	10.7	11.1	9.6
Protein, Kjeldahl, nitrogen × 6.25	34.1	40.6	7.4
Fat, Soxhlet	4.8	7.3	1.0
Ash, Calcination	3.3	3.4	2.7
Fiber	8.6	1.7	46.0
Extract, nonnitrogenous	28.5	35.9	33.3
Calories, kilocalorie per l00 gram	290	366	174

tribution by the symbiotic action. Proof of this is that 160 kilograms per hectare (143 pounds per acre) of ammonium nitrate was applied experimentally without producing any response. The soil had good conservation management with 5% organic matter and high phosphorus levels.

The broadleaf and grass weeds were chemically controlled using preemergence Simanex applied at a rate of 2 liters per hectare (Agan Chemical Manufacturers Ltd.; see Section 2.8.4).

Grass weeds in the lupin crop were controlled with Poast (Sethoxydim) applied at a rate of two liters per hectare. The post-emergence broadleaf weed control is difficult because adequate herbicides are not available in the market. For that reason, good pre-planting and preemergence weed control are very important.

The narrowleaf lupin (*L. angustifolius*) is shorter than the *L. albus* and *L. luteus* cultivars and has an ascending inflorescence, forming from two to four strong peduncles with dehiscent pods that do not twist easily. This prevents major losses by shattering, achieving up to seven seeds per pod. Its physiological maturity is 30 days earlier than the other varieties, reaching maturity by mid-December at Chequén. When the leaves fall, the stem remains light green and the seeds maintain 16% moisture. The mechanical threshing is performed without problems, but some losses by fallen pods occur if the correct time for the threshing has passed; however, the resistance to shattering is greater, both in preharvest and harvest, than other lupin varieties.

According to the analysis in Table 1–6, the lupin seed has excellent advantages for hog and poultry feed, especially regarding the protein and energy value. The advantages are even more when it has been shelled. The hull can be used for cattle feed complemented with other by-products or grains. T.P. Mironova and others state that the grains have from 30 to 36% crude protein, from which 82 to 94% is composed of albumens and globulin. This gives it high nutritional qualities. Also, the lack of trypsines and chemiotrypsines inhibitors improves the quality of this grain in relation to other legumes (Mironova et al., 1990).

We have not seen any foliar diseases during the vegetative period with the exception of the brown spot disease. This indicates some resistance to the common diseases.

The brown spot disease, in which the seeds of affected plants show a coffee color, is caused by the imperfect fungus (*Pleiocheta setosa*). This fungus is transmitted by infected seeds or through the crop residue, and it stays active for 2 years (Lara and Andrade, 1983). This disease can be controlled if Sportak 40EC (Schering Agrochemicals Ltd., England) is applied at a rate of 1.0 to 1.5 liters per hectare during the winter.

At Chequén, this fungal disease affected a small area of the crop in the second crop rotation year with wheat. The infected plants were found to be poorly nourished as a result of the soil's excessively coarse texture. Other planted areas were not affected.

According to this limited experience, the brown spot disease attacks the plants severely when adequate humidity and fertility conditions do not exist. The phytopathologic analysis of the seeds of this lupin variety did not show the presence of fungus *Pleiocheta*, but did show small quantities of *Penicillium* sp. and *Mucor* sp., both are known as an environmental fungus.

Sweet lupin affected by the brown spot disease (1990).   Sweet lupin affected by anthracnose (1990).

The bitter blue lupin is an excellent soil improver (1991).

Anthracnose caused by the *Colletotrichum gloeosporoides* fungus. During 3 years of planting this variety, there was no significant damage from this disease. Sportak 40EC applied during the spring at a rate of 1.0 to 1.5 liters per hectare can control this fungus. The seed should always be disinfected with a proper fungicide that does not inhibit the plants inoculation with nitrogen-fixing bacteria.

Due to this legume's excellent botanic and agronomic characteristics, I can say that its future is good in the soils of the Coastal Mountain Range (Regions V to X). This legume is easy to manage, and optimum results will encourage thousands of farmers to produce a high protein value legume seed. This lupin can be a substitute for imported soybeans, cotton, peanuts, or other high protein meal by-products. It can be used to feed poultry, hogs, and rabbits as well as humans. Slug control must be very effective to avoid damage to the crop.

Lupin yields can be higher than 4.0 tonnes per hectare (3572 pounds per acre). In 1988, the first crop year at Chequén, the yield was 3.5 tonnes per hectare (3125 pounds per acre), which has increased slightly in the last 2 years, mainly because of a better weed and slug control.

The most reasonable way to plant this grain legume is no-till, which allows the producer not only to obtain a good income, but also to achieve excellent free benefits in soil conservation. Although the seeds are broadcast over the soil surface, they are not eaten by most birds. The ideal is that they be partially buried or protected with some crop residue. Inoculation is not needed if a lupin variety has been planted previously or if the traro shrub (*L. microcarpus*) grows on the site.

1.4.2.3.3. Blue Lupin (*Lupinus angustifolius* L.)

The blue lupin (*L. angustifolius*) was introduced in Chile in 1910 with the purpose of incorporating it as green manure to improve the organic matter levels and soil fertility. This lupin adapts very well to extremely infertile soils. It is capable of improving the soil productivity at Chequén. To achieve this, simply spread 80 kilograms per hectare (72 pounds per acre) of seed over any soil without cultivation, whether the seed is covered or not. North of Region VIII, planting using a light disk harrow to bury the seeds can reduce seedling losses caused by the lack of moisture. In the erodible soils of Regions VI to X planting should be performed n May and June.

At Chequén we are using blue lupin as a green manure and residue covercrop as well as for seed production.

The plant density of blue lupin is low during the crop's first year. This situation is reversed in the next year, and high quantities (up to 30 tonnes per hectare; 13 tons per acre) of green biomass are generated. To achieve full soil fertility and conservation benefit, the second year crop should be cut with any type of mower. The mowing should take place between the end of flowering and the beginning of pod formation. This mowing also will cut the weeds before their seeds are formed. After this weed control, care also should be taken to control weed and lupin regrowth so they do not invade the field again. A field with these conditions is ideal for no-till planting.

The blue lupin resists the most common diseases that affect these legumes; however, after 3 years of being established as a covercrop, the brown spot disease and other phytosanitary problems destroy it. This limitation does not affect its benefit as an improver of soil fertility. This objective is obtained during the second year. The alkaloid content in the seeds and plant makes them unsuitable for animal consumption.

As a legume plant, the blue lupin can fix atmospheric nitrogen into the soil. This forms many active nodules on the root system and allows a savings in nitrogen fertilizers.

The blue lupin produces good biomass with minimum nutritional requirements. At Chequén, the plant does not show responses to monocalcium phosphate fertilization nor to potassium applications, but it is very important to take care of the soil's pH. This plant shows good responses to lime (calcium carbonate) applications where the pH is lower than 6.0.

LUPINO ANGOSTIFOLIUS
AMARGO

LUPINO ANGOSTIFOLIUS
DULCE

LUPINO MICROCARPUS
YERBA DEL TRARO

LUPINO LUTEUS
VAR. AUREA BAER

These lupin seed varieties are planted at Chequén with the exception of the traro (yerba del traro) forb, which grows native in the area. Please note that *L. luteus* var. aurea Baer has been mislabeled and should be labeled as *L. albus* var. *Multolupa*.

This legume has a deep taproot with multiple side branched roots and abundant nodulation. This characteristic has a positive action over the soil's physical condition, loosening very compacted areas and letting the old root space be used by future plants. This improves the water-holding capacity and water infiltration, reduces the soil loss by erosion, and improves plant available moisture. Also, any crop in rotation with the lupin is benefited. The yields are higher as a result of better nutrition, especially due to high levels of nitrogen and phosphorus in the soil, and also because the large plant density and vegetative biomass production help to control weeds.

Although lupins are a winter crop at Chequén, the planting of the sweet lupin (*L. angustifolius*) in the spring has been successful. It had a yield of 3300 kilograms per hectare (2950 pounds per acre). This is good, but can be improved by better weed control and reduced threshing losses

The traro forb, a native legume plant, grows in the dry soils in the Coastal Mountain Range, Cauquenes County (1988).

by shattering and pod drop. The plants were stronger and produced pods in abundance, maybe because they were sprinkler irrigated in December and January.

The lupins, whatever the variety, should be considered for inclusion in the winter rotation. The undeniable benefits of high yields, soil improvement, and conservation make lupins an important crop for poor and degraded soils.

If lupins are planted in the fall, this should be done after the first rainfall. The annual weeds should be controlled as soon as possible before planting.

The extraordinary adaptation of this legume is caused by the unnoticed presence of the forb Traro (*L. microcarpus*). This wild plant is native in most of the soils from Atacama to Patagonia. It keeps the rhizobium bacteria active, naturally inoculating all the lupin varieties. This fact is very important for the restoration of fertility and organic matter levels in the poor and degraded soils in south-central Chile.

A sour taste attributed to the lupin varieties is caused by nonprotein nitrogen alkaloids that are toxic. These alkaloids are soluble in water, which can aid in consumption of treated grain. Certified seeds of sweet lupin, a product of genetic research, produce levels that do not exceed 0.05% of alkaloids. This low percentage is safe for animal consumption (Baer, 1986).

1.4.2.4. Oilseed Rape or Canola (Brassica napus L.)

This oil seed plant belongs to the cruciferous family. It has a special value for the no-till system. Some canola varieties grow well in clayey soils under no-till. Because the seeds are very small, it is important to maintain the planting depth and seeding rate.

Another important item in canola establishment is the sensitivity to allelopathy effects produced by the residue of small grains. Small grain straw must be baled or windrowed when it is more than 3 tonnes per hectare (2700 pounds per acre). In some areas, slugs can be a serious prob-

Oilseed rape (canola) seeds (1991).

Canola in wheat stubble, Montpellier Farm, (Mulchén County, 1990).

Canola in wheat stubble (1991).

Canola in wheat stubble, Las Bandurrias Farm (San Fabián County, 1988).

lem with no-till. In no-till, canola plant roots have a tremendous impact on the soil physical conditions. Because of the improved soil porosity the water holding capacity of the soil is increased. The following crop in the rotation (wheat) responds well to canola residue.

The establishment of canola with the no-till system has been a success in sandy soils (trumaos) of the lands close to the mountain range. Canola makes an excellent rotational crop for grains. The deep taproot of this plant is similar to that of the lupins, which benefits the physical characteristics of the soil. At Chequén, we have found canola roots that are 2 meters (6 feet) deep.

It is important to point out the amount of canola residue produced by the crop. Although the volumetric quantity of organic material appears to be high and possibly affects the upcoming crops, canola residue can be chopped very easily with a forage chopper or the harvester combine (as shown in Section 2.5.6). Like other broadleaf plants, the canola must be planted in fields free of broadleaf weeds because only a few herbicides control these weeds. Grass weeds are easily controlled with postemergence selective systemic herbicides (see more information in Sections 2.8.3.4. and 2.8.3.5.).

1.4.2.5. Potatoes (Solanum tuberosum L.)

Although potatoes grow beneath the soil surface, their behavior in a no-till system was unexpected. At Chequén, we planted potatoes in corn residue on top of the soil with a prior application of ammonium nitrate and glyphosate. The rows were planted 70 centimeters (28 inches) apart and then covered with a mixture of semi-decomposed straw and pine sawdust. The planting was completed at the end of winter and was irrigated by sprinkler.

We observed that the potatoes developed normally in the organic residue cover, introducing part of their roots into the soil. Some potatoes were left partially buried. We observed the curious

Potatoes to be harvested were planted by hand in corn stubble and covered with straw and sawdust.

A reseeded permanent pasture under intensive use. Electric fences are used to control livestock movement (1978).

phenomena of roots penetrating corncobs, developing the fruit on the other side. Potatoes introduced in corn stalks developed clean and abundant fruits with a delicious taste.

I believe that this experience can be very productive in more humid soils than those of Chequén. In Chiloé Island, Region X, some farmers plant potatoes without tilling the soil, only covering them with seaweed and semi-decomposed grass. The results are profitable, erosion is controlled, and soil fertility is maintained.

1.4.3. Reseeded Pastures

The practice of regenerating degraded pastures was initiated in Chile in the 1960s. A meadow recovered without tillage or cultivation and with low planting costs also benefited the soil. One reason that the livestock producers support this practice is the possibility of grazing the land early after reseeding. This is because regeneration of pasture does not involve soil movement, thus avoiding damage caused by the cattle trampling in rainy seasons.

Forage seeds generally are well adapted to no-till; however, some legumes such as alfalfa (*Medicago sativa* L.), need to have the seed with good soil contact in the furrow and have adequate planting depth and soil moisture.

An electric fence facilitates better use of direct grazing. It is considered a valuable tool in the pasture and livestock management.

Generally, it is suggested to increase the seeding rate by 10 to 20% in no-till management because the percentage of seed loss is higher than with traditional planting. This is especially valid when the planting is done in fine-textured soils that have a high moisture content. The seed zone can lose its humidity if the furrows are not closed. Under some conditions the planting depth is not sufficient to cover the seeds, resulting in poor germination.

Weed control in a fruit orchard using horizontal rotary chopper (California, USA, 1958).

1.4.4. Orchards and Vineyards

The first trials in modern times to manage the soil without tillage were done in orchards and vineyards. In 1958, I visited vineyards in California that were planted without tillage and were producing excellent yields. Weed control was performed by residue choppers between the rows. The weeds, chopped several times during spring and summer, were left over the soil, forming a useful mulch cover. This mulch helped control the germinating weeds. For that reason it was necessary to introduce a winter green covercrop (oat and vetch) in the following years to keep the soil more biologically active and to control the moisture lost in spring and summer due to direct soil evaporation.

Today, orchard and vineyard management does not use the traditional tillage and planting tools because it has been found that erosion is reduced, costs are reduced, weed control is improved, and rainwater or irrigation water is more efficiently used. It also is important to consider that superficial roots are not damaged by harrows or other tillage implements.

The vineyard no-till system was started in Chile around 1958 on the San José Farm (Ránquil County, Region VIII) owned by Alberto Grüebler and his son. Today most of these vineyards are maintained without cultivation and with excellent production and quality. The results of this practice show that after 14 years of conservation management, the level of organic matter in the upper 5 centimeters (2 inches) of soil increased from 1.7% with cultivation to 3.42% with a no-till system. In the next soil horizon to a depth of 30 centimeters (12 inches) the increase was 50%. All this was attributed to the leaves and chopped weeds accumulated on the soil surface (Riquelme et al., 1972).

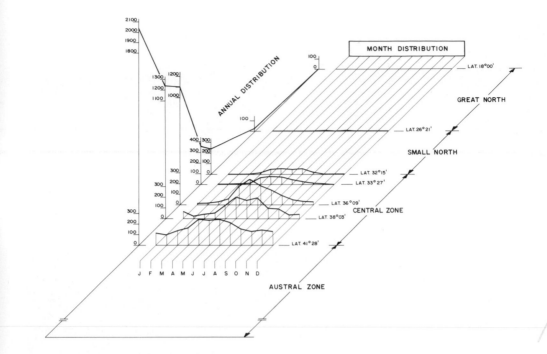

Fig. 9. The precipitation distribution in continental Chile (Raggi, 1989).

No-till management of orchards and vineyards is simpler than that of annually planted land, mainly because mechanical weed control is easier. In some cases, difficult weeds must be controlled. The application of contact, systemic, or residual systemic herbicides complements weed control.

1.5. No-till on Dry Land and Irrigated Land

1.5.1. Dry Land

South-central Chile is known for high rainfall rates in the fall and winter (April through September), when 70% of the annual precipitation occurs. In the province of Concepción (Region VIII) the average daily precipitation during this season is more than 10 millimeters (0.4 inches) on 39% of the rainy days. In relation to intensity, 18% of the events are at a rate of more than 10 millimeters in 60 minutes (Endlicher, 1988). During spring, the precipitation is not enough for crop production. The summer is characterized by a strong subtropical insolation. This condition is suitable for higher production when the soils and crops are efficiently irrigated and managed. Figure 9 shows the precipitation distribution table in continental Chile.

According to this table, the zone with highest precipitation is the area close to the mountains. Farming, especially grain production, is performed in dry barren soils. The coastal dry barren soils have been under intensive cultivation for decades, maybe centuries. This has resulted in severe soil erosion of the upper soil horizons. Consequently, the reduction of natural fertility was severe.

During the colonial era of the 1800s, the irrigated land in the valleys of the country was limited. Farming was performed basically in rainfed soils. The cultivation of these soils began in the 18th century, reaching its height in the second half of the 19th century. During that time, especially during the gold rush fever in the California region, rainfed soils of the coastal mountains were high wheat producers. Grain was exported to other countries. The strong demand for this grain expanded the areas planted from the central to the southern part of the country.

The rainfed, coastal and interior wheat lands form a long, narrow belt. The oldest soils under cultivation are those in Region V (Valparaíso). As these fields lost their productive capacity, they were abandoned, and the farming operations were moved to more fertile soil zones in south-central Chile. Wheat is currently planted from Region VIII (Chillán) up to Region X (Osorno).

Further south wheat is planted, where the winter rains are intense and, consequently, the erosion problems are more severe. The erosion process is alarming in this area where wheat is planted on steeper soils.

In relation to this situation, Elíias Letelier, an agronomist wrote:

> "The most visible and important negative effect in a long run with the excessive winter rainfall in Chile is without any doubt water erosion. Wheat cultivation is the main cause of this type of erosion and also the crop that suffers the most consequences from erosion. Most of the soils with slopes between 5 and 30% are very susceptible to erosion, but still used for this crop" (Letelier, 1988).

In Chile, the geographic limit of grain production is Region X. The climate in the southern region is not adequate for grain production because of excessive rains and the lack of temperature. The degradation of the coastal mountain landscape and the erosion of the soil are clear consequences of the bad use of this fundamental natural resource. The soil loss in some areas is higher than 1 meter (39 inches). This means that horizon A is lost and horizon B is seriously threatened. Big gullies make these areas unproductive. The rough landscape will remain like this for centuries or until it reaches a natural equilibrium.

The pine tree reforestation in Chile currently covers 1,300,000 hectares (3,211,000 acres). This reforestation is an extraordinary alternative for the protection of these eroded soils. Tree establishment is easily achieved, and adaptation to different soils is exceptional because of the favorable agroclimatic conditions. Without doubt, the reforestation of this dry, barren area constitutes the most natural and efficient way to conserve soil and water on steep surfaces where the winter precipitation is high. At Chequén, we observed that no immediate runoff occurs during intensive rainfall on the 20-year-old plantations. This indicates that the soil cover is more efficient than those originated by the meadows and no-till system. The pine trees, with their particular foliage and needle-shaped leaves, intercept part of the precipitation. Some of the water that does not infiltrate the soil is evaporated to the atmosphere directly from the pine foliage and soil surface. Although this fact can limit their planting in some areas, **the only way to make this eroded soil profitable is with reforestation**. In these dry, barren areas pine trees are easily established, without development limitations, and constitute the only conservation and economic alternative. Because of this I can say that **woodland, natural or planted by humans, is the most perfect way of no-till**.

The establishment of forage species in dry, barren meadows, whether they are natural, improved, or artificially introduced, has been a great conservation value and, at the same time, has produced higher incomes.

If we insist in planting grains on these eroded soils, soil tillage implements must not be used. The establishment of grain by no-till makes it possible to grow crops on soils that are susceptible to erosion. The Chequén Farm is in the center of the erosion belt in Chile, but no-till management is used and higher yields are produced. At the same time conservation and improvement of the soil are excellent.

The soils of Chequén have shown an extraordinary adaptation to the no-till system. Although the conditions have not been adequate, the results are good if the soils are well managed, which obviously means a greater need for applied technology.

The Coastal Mountain range has been extremely altered by inappropriate land use (Florida County, 1986).

Reforestation of eroded land with pine trees resulted in control of soil erosion, improved soil fertility, and a better Chequén economy. The gully reforestation controls soil erosion and at the same time is the only feasible economic alternative.

The lack of fertility and, even more, organic matter, does not disqualify eroded soils from the establishment of pastures or no-till farming. With efficient fertilization and good weed control, the initial yields should be acceptable. Nitrogen fertilization is important in the coastal dry, barren soils because the excess residue from small grain crops can cause some problems with planting, especially when planting should be completed immediately after the fall rains begin.

The south-central Andes foothills have light soils of volcanic origin. These soils are characterized by incipient profile development and are taxonomically classified as Andisols. They are exceptionally well adapted to no-till farming, but phosphorus fixation can limit their use. This problem is caused by aluminum and iron in solution and, depending on the soil pH, the amorphous mineral called allophane (Besoain, 1958; Espinos, 1968; Schenkel and Bleiholder, 1967). To overcome this problem, phosphorus fertility must be reinforced during the first 4 to 6 years of crop production. These soils are difficult to manage because of their chemical or fertility deficiencies. Nevertheless, they are the best ones adapted to no-till because of their excellent physical characteristics.

1.5.2. Irrigated Land

In Chile, the potential area for irrigation is 2.5 million hectares (6.2 million acres), on which 46% has permanent irrigation systems with 85% guaranteed water; 32% need supplemental irrigation; and 22% is dry land where irrigation must be used. Of all the gravitational irrigation methods, the most common in Chile is the basin or flooding system. It is used in more than 70% of the irrigated area, with a 15 to 30% efficiency (Raggi, 1989).

Only a few farmers practice no-till with irrigation in this country, so there is almost no experience in this situation. We may think that irrigation can be difficult if the soil is not plowed. Irrigation of summer crops is generally performed by the furrows between rows, which are constructed using the loose soil after planting the crop. However, a border irrigation system permits the establishment of the no-till crop without difficulties. Land leveling is needed to simultaneously build the respective borders. The width between the small dikes depends on the water amount that will be available and the length, slope, and texture of the soil to be irrigated. In some circumstances, the stubble on the soil surface reduces the water velocity, which has to be considered in determining the length of the borders.

A sprinkler irrigation system is the most adapted to no-till systems. It simulates the natural rain effect, obtaining a higher efficiency in water use. It also allows planting on soils that have a wide range of slope without erosion risks (Soza, 1980).

The steep topography of Chequén has restricted establishing an irrigation system other than a sprinkler system. Sprinkler irrigation produces high corn yields on Class IV and VI soils. No-till planting was initiated on Chequén with corn irrigated with sprinklers, and the results are better every year. Annually, we plant 30 hectares (75 acres) in corn on sloping lands, some of which are very steep (35%). We do not irrigate wheat when it is planted in the fall following the crop rotation system.

It is important to consider that when using no-till management, the soils are not disturbed, so irrigation furrows cannot be constructed. Where sprinkler irrigation cannot be used, it would be adequate to irrigate by borders. In this case the stubble should be left evenly spread and the water slowly moved around the soil and residue until the edge of the border is reached. A lower water requirement is anticipated with this system because the residue on the soil reduces moisture loss by evaporation. Also, steeper soils can be irrigated with less risk of erosion.

*Standing proud above the
canopy of your peers,
you enchant the disbeliever,
splendid green mantle.
You cover my hills,
you hide my cracks and miseries
and convert your squalid soul
into a generous heart.*

*More of a patriot than a foreigner,
notable is your fertile labor,
you tear away our misery
and bestow us with wealth.*

Medium sized sprinkler irrigation on no-till corn (1990).

Sprinkler irrigation on lupins (1989).

High pressure sprinkler irrigation on no-till corn (1989).

Sprinkler irrigation on areas of steep slopes (1989).

Center pivot irrigation equipment in central Oregon, USA (1986).

Linear sideroll irrigation for soils that are flat and crops that are short, central Oregon, USA (1986).

This system of border irrigation on pasture can be a good alternative for irrigating corn or other crops planted by no-tillage, Oregon, USA (1958).

1.6. THE MECHANIZATION IN NO-TILL

To have success in the management of no-till, the proper equipment is needed to do the work. Farmers think that it requires a big investment in equipment to change over to the no-till system. The truth is that you can start practicing no-till with conventional drills and planters if the soils are medium to coarse-textured and if very little residue is on the soil surface.

A fundamental condition that must be observed is the inclusion of crop residue in the management of this system Burning or complete removal of straw from the field must be eliminated. This is the key to agronomic success in the future. It is important to have a no-till planter capable of doing the best possible job under most soil conditions, and adequate amounts of crop residue must be left on the soil surface.

Fields that are in continuous cultivation will certainly respond to equipment that can be used in conservation treatment of the soils. It is possible that one can change a conventional planter to a no-till planter and probably do away with many of the field implements, such as the plow, disk, field cultivator, and chisel plow.

If the first efforts are to no-till into crop residue of corn, wheat, canola, soybeans, lupins, or other crops, then it is necessary to efficiently manage the equipment and residue of the crop that precedes the planting.

1.6.1. Tractors

Tractors in the management of no-till should have different characteristics from those used in conventional tillage.

No-till tractors should have:

1. High flotation, low pressure tires
2. Four-wheel drive
3. In general, maximum of 140 horse power

No-till tractors should not have:

1. Water added to tires as counter-weight
2. Counter-weight added to the rear axle

Generally speaking, no-till tractors must be lighter than conventional tractors. With the need for less traction in no-till, the power of the tractor can be directed more toward the power take-off, for example, as used in chopping residue. Tractors used in no-till should also have more power per unit of weight.

1.6.2. Combines and Harvesters

No-till must be carried out on the residue from the preceding crop. To be effective, the grain combine must be equipped with a straw chopper and spreader and must distribute the residue uniformly. This equipment is generally optional on combines; however, the cost is easily justified by the job it does to simultaneously harvest the crop and then chop and spread the residue over the same width of the harvest. Also, the chaff and short residue as well as the large pieces of residue must be spread evenly.

Even though equipment manufacturers do not provide power drives for fans, low cost fans that use forced air behind the straw walkers are available. This air is blown by two nozzles across the back of the grain separator. This can eliminate the allelopathic effects (from the residue of small grains) caused by the high concentration and poor distribution of the fine straw particles. Thus, this residue can be distributed over the soil surface, and crops can be planted directly into it. Caution must be used to not produce a negative allelopathic effect and/or an unbalanced condition between carbon and nitrogen.

High flotation tires on combines and harvesters reduce pressure on the soil, thus reducing soil compaction.

1.6.3. Stalk Choppers

If the combine does not have a way to chop and redistribute the residue, then another machine must be used to obtain the same results. The most common choppers available are forage harvesters and horizontal rotary shredders.

1.6.3.1. Choppers or Forage Harvesters

This machine is, in general, the most useful on the farm. It is an efficient mower that can cut and load forage onto a wagon in one trip across the field by using the power-take off (PTO). It can be used for chopping thick stalks, like corn, sunflower, and canola.

The technical concept is that the chopper works by lifting and cutting the forage. A horizontal axle carries multiple knives spinning in a vertical motion that produces the vacuum to lift and cut the forage at the same time. This action also can lift and chop crop residue. Therefore, this machine also can be used as a regular residue chopper. To make this work, the forage chute must be removed and the back end of the chopper must be open to permit an even distribution of the residue.

Self-propelled combine equipped with a straw chopper and spreader (1995).

Self-propelled combine with a spreader system to redistribute the straw (Chillán County, 1989, Courtesy of Agronomist Pedro del Canto).

Self-propelled combine harvesting 8000 kilograms per hectare (119 bushels per acre) wheat. Notice the high concentration of straw leaving the rear of the machine (Lautaro County, 1989).

A modified chopper used for cutting and shredding the thick stalks of such crops as corn, sunflower, and canola (1989).

Chopper for harvesting forage prior to planting with no-tillage (1990).

Horizontal shredder (rotary mower) for shredding and redistributing fine straw material (1988).

Commercial residue choppers are available that can be used in any kind of residue. These machines are built as a three-point hitch and a pull type model. They may be used just before planting with a no-till drill. This type of machinery is one of the most important on the no-till farm.

1.6.3.2. Horizontal Rotary Shredder

This machine has two or more horizontal knives that are moved by the PTO. These are efficient mowers, choppers, and spreaders of thin straw, such as wheat, oats, barley, and rye. The chopped residue is placed only over the area where the equipment is driven.

This equipment generally does not have the lateral opening necessary in either direction to allow the residue to flow easily from the machinery. It is necessary to cut an opening for an easy exit if this has not been done in the factory. This equipment is available with a hydraulic three-point hitch, as an integral pull type, or with the shredder attached to the frame of the tractor. The use of one or the other types depends on various factors and preferences of the farmer.

1.6.4. Residue Mowers

1.6.4.1. Sickle Bar Mower

The sickle bar mower has a simple design based on an alternating, oscillating set of knives that can be powered by mechanical traction, pulled by animals, or by the PTO of a tractor. In working with crop residue, the important thing to remember is that the equipment only cuts the residue, but does not spread it; however, it can be used in the control of weeds, as discussed further in Sections 1.6.4.2 and 2.8.6.

1.6.4.2. Disk Mower

These effective mowing systems have three or more disks that turn rapidly in a horizontal direction. The disks have knives attached to the edges that permit cutting without plugging. This mower is powered by the PTO of the tractor and can be pulled as an integral hitch or connected to the hydraulic three-point hitch of the tractor. Whatever the job to be done, this type of mower is more efficient and faster than the sickle bar mower yet produces the same results.

This mower does not have a major use for crop residue, yet it can cut whatever straw residue that is left after the harvesting and baling operation; however, if weeds are growing, the mower can be beneficial in preventing the weeds from going to seed.

1.6.5. Balers

In some circumstances, too much residue can cause problems. This is especially true for rye, wheat, triticale (× *Triticosecale rimpaue* Wittm.), barley, and oat residue, which can exhibit allelopathic effects. We will see in the following chapters that this will probably require the removal of a part of the residue. For this condition, whatever type of baler, rectangular or round bales, can be used to reduce the amount of residue left in the field. If the straw is to be baled, the combine can leave the straw in a windrow, considering the uniform distribution of the small chaff as mentioned in Section 1.6.1.

1.6.6. Side Delivery Rake

This low cost implement is very useful working with small grain straw in no-till. It works fine with a small tractor, three-point hitch, or pull type. The most adaptable side delivery rake for proper straw management has three or more steel wheel reels.

This implement can help farmers make the necessary windrows on cereal straw if the next crop will be drilled in the fall. The straw is raked into windrows that are widely spaced to permit planting, spraying, and harvesting in-between. This management system eliminates the need for fire or baling the straw, which helps avoid the negative allelophathic activity and nitrogen immobilization during the next crop growing period. The width of the windrow must be the width of the herbicide applicator boom, leaving the straw in narrow parallel windrows. After crop harvest, these windrows are redistributed over the soil surface using the rake or a residue chopper.

1.6.7. Herbicide Applicators

1.6.7.1. Pressure Sprayers

Without doubt, a pressure sprayer is one of the most important tools in the management of no-till and is effective when properly used. The pressure sprayer hooks to the tractor by the three-point hitch or it is pulled by an integral hitch at the drawbar of the tractor. The control of weeds is fundamental to obtain success in the harvest, especially in the application of preplant herbicides. Good control of the weeds is obtained if herbicide is uniformly applied for total coverage of the surface to be treated. Stakes or flags should mark the path of the sprayer. To have overlap, the markers should be 50 centimeters (20 inches) less than the width of the boom. This overlap is very important, especially in hilly areas, to avoid skipping areas or double coverage. Foam markers are available that guide the operator and help avoid sprayer skips.

This equipment works properly if it has the following features.

- Uniform flow pump (2 or 3 diaphragms)
- Mesh filter that eliminates large particles in the water
- An assortment of sprayer nozzle tips that permit different flow rates
- Sprayer nozzles that are easily cleaned and have individual filters coupled to an anti-leak nozzle system
- High and low pressure gauge
- Liquid level indicator in the tank
- System to completely empty the tank
- Boom width and tank capacity that is suitable to the topography and area to be treated
- Individual valves on the boom that permit each section of the boom to work separately
- Liquid tank agitator
- Easy access to tank filling
- Construction with anti-corrosive material

Spraying equipment that operates at a faster speed and in windy conditions is now available. This equipment functions on a current of air that is produced by a fan, channeled by a hose, and covers the total length of the sprayer boom. This equipment also permits a better contact of the chemical mixture with the leaves and stems of the weeds. Equipment that monitors the area and speed, while automatically regulating the predetermined flow rate is available for these sprayers.

Round baler.

Round bales are ideal for feeding animals grazing in the same field (Mulchén County, 1989).

Rectangular baler baling a windrow of straw left by the combine (1990).

Side delivery rake is adjusted to move the cut hay to improve drying before baling (1990).

Pressure sprayer mounted on a hydraulic boom in front of the tractor (1991).

Pressure sprayer developed by ICAT-INIA is powered by animal traction. It has a 6 meter (19-foot 8-inch) boom (Cauquenes County, 1990).

Pressure sprayer with an 8 meter (26 foot 3 inch) boom mounted on the rear of the spray tank (1988).

Front mounted pressure sprayer with plant shields, designed and constructed in Chequén, applying paraquat for control of annual grasses in corn (1991).

1.6.7.2. Controlled Droplet Applicator

The controlled droplet applicator (CDA), is based on the development of small droplets that fluctuate between 70 and 300 microns (a micron is equal to one thousandth of a millimeter) with which low volumes of water, 20 to 80 liters per hectare (2 to 8.5 gallons per acre) can be used. This equipment, driven by a small electric motor, spins the chemical liquid past a toothed disk and breaks the liquid into very small droplets. Depending on the size of the droplets that one wants to apply, the spinning disk can spin between 2000 and 5000 revolutions per minute.

In our 3-year experience using this equipment on Chequén, we did not obtain good results. The principle problem observed in the use of the CDA was the lack of thorough wetting of the plant stems and leaves. This became serious with systemic and contact herbicides. The preplant control of actively growing weeds with systemic herbicides was insufficient, and a high density of plants that totally covered the soil did not permit sufficient foliar wetting.

This equipment, which reduces the amount of water used per acre, produces a high concentration of herbicide, which in many cases can be toxic for the operator who breathes the contaminated air. The author has had some unpleasant personal experiences with paraquat using this system.

We tested different sprayers and nozzles at Chequén. After 15 years of using this equipment, we concluded that in no-tillage it is important to use the highest level of water recommended for the chemical product. This is especially true when using systemic and contact herbicides. For actively growing and dense weeds, glyphosate has a better impact when it is applied in 350 liters of water per hectare (37 gallons per acre) compared with 200 liters (21 gallons).

Applying limestone with a pendulum spreader system (1990).

The Argentinean Association of No-till Producers has observed that the application of Sencorex (metribuzin), at 130 liters of water per hectare (14 gallons per acre) produced poor results. Changing to 250 liters of water per hectare (27 gallons per acre) gave excellent control (Aapresid, 1990).

1.6.8. Fertilizer Applicators

Fertilizer applicators are efficient machines. The drop or broadcast spinner evenly applies granular fertilizers to fields in different doses after planting. To obtain a uniform distribution, the application rows need to be marked with stakes or other markers at the correct width of fertilizer application. In flat terrain, the operator can be guided by a field helper with flags. The helper stands at the extreme edges of the field. This equipment should not be used when the soil is saturated with water. Excess compaction can result from the weight of the tractor and fertilizer spreader and can damage the seedbed, the soil, or both.

1.6.9. Lime Spreaders

Correcting the acid condition in the soil is important. An effective way to correct acidity is to apply calcium carbonate (a chemical salt that is present in a fine powder form). This compound should not be broadcast during high winds because the material lost can result in poor application.

Lime can be spread using either a pendulum or drop spreader. A pendulum spreader disperses the lime by single or double paddles situated inside the box. The lime is spread over narrow lines. This old pull type spreader is very useful in windy climates. The spinning broadcast spreaders of granular fertilizer have a shorter pendulum stirrer and place an agitator in the bottom of the cone. Another type of spreader that has a much higher load capacity is a wagon with twin disks in the rear section. It does a good job even with powdered fertilizer. This spreader applies lime faster, but is not as efficient as the others mentioned.

1.6.10. No-till Planters

This machine made seeding without prior soil tillage possible. A slit is made by a disk in front followed by two disks that simultaneously plant the seed and fertilizer. This is accomplished in the row made by the front disk, which cuts the crop residue and roots, forming a narrow seedbed that will facilitate the development of the roots and plants.

This disk can be smooth, serrated, rippled, or fluted. The smooth disk is used for seeding small grains, for example, wheat. It is sometimes on no-till planters equipped with three disks: a front coulter followed by two disks attached to the seeder and fertilizer applicator. The fluted disk is used with many soil conditions. It is recommended by the manufacturer for the seeding of large grain, for example, corn.

At Chequén, we designed a front cutting disk that uses a smooth, serrated disk. This design was tested and demonstrated to have the highest capacity to penetrate through heavy residue such as corn, wheat, and lupins. The disks have sufficient ability to cut the residue and roots and at the same time can properly maintain the correct depth for seeding. This planter requires 80 kilograms (176 pounds) of weight over each coulter unit for planting small grains and 150 kilograms (331 pounds) to plant large grains.

Our experience shows that the smooth, serrated coulter is more efficient than the rippled or fluted coulters. In reality, all that the seed needs for good germination is a slit in the soil. The smooth coulter can accomplish this efficiently when the quantity of the residue is like that of corn;

Liming with a drop spreader (1989).

Lime deposited by a drop spreader (1989).

(1) Serrated disk, 38 centimeters (15 inches) in diameter, with each tooth indented 2.5 centimeters (1 inch) for cutting thick stalks such as those of corn, sunflowers, and canola. (2) smooth disk, 38 centimeters (15 inches) in diameter for cutting fine straw and thick roots. (3) Serrated disk, 38 centimeters (15 inches) in diameter, with teeth indented 20 millimeters (3/4 inch) for cutting more than 3 tonnes per hectare (2700 pounds per acre) of straw.

(4) Ripple coulter, 38 centimeters (15 inches) in diameter, for soils with less than 3 tonnes per hectare (2700 pounds per acre) straw. (5) Fluted coulter, 46 centimeters (18 inches) in diameter and 5 centimeters (2 inches) in width, for planting large seeded crops in less than 3 tonnes per hectare (2700 pounds) per acre of straw. (6) Bubble coulter 38 centimeters (15 inches) in diameter, for multiple use; especially suited to deep seed placement and wide rows (1990).

(7) Fluted coulter, 43 centimeters (17 inches) in diameter and 2.5 centimeters (1 inch) in width, designed for planting soybeans and intermediate sized grain in fine textured soils. (8) Serrated coulter, 46 centimeters (18 inches) in diameter, with 20 millimeters (3/4-inch) indentations for cutting fine stemmed residue over 3 tonnes per hectare (2700 pounds per acre; 1990).

A no-till planter unit made by Semeato in Brazil. The coulter, parallel seed unit, fertilizer unit, and press wheel are shown (1990).

A no-till planter made by Semeato in Brazil. This planter can position the front coulter to have better cutting and soil penetration because the disks are offset (1989).

A combination no-till drill, Juber 2000 from Argentina, with a double disk opener of 37.5 centimeters (15 inches) in diameter (Courtesy of Agronomist José Astudillo, 1990).

A unit from the John Deere no-till drill Model 750 from the USA. It has only one disk of 46 centimeters (18 inches) in diameter inclined at 7 degrees.

No-till drill with fertilizer attachment made by John Deere Company, USA. The Model 750 has 16 rows separated at 19 centimeters (7 inches), and a total width of 3.05 meters (10 feet) Courtesy of Agronomist José Astudillo 1990).

No-till drill with fertilizer attachment made by Semeato in Brazil. Model TD 400 has 24 rows each separated at 15.8 centimeters (6 1/4 inches) and a total width of 4 meters (13 feet; 1989).

No-till drill with fertilizer attachment made by Bettinson in England. Model 3-D equipped with smooth front disks of 33 centimeters (13 inches) in diameter. This company was a pioneer in the introduction of small grain no-till drills (Oregon, 1986).

No-till drill made by Tye Company, USA. This model, Pasture Pleasure, is equipped with 43 centimeters (17 inches) in diameter ripple coulters on the front. It can plant 10 rows spaced at 20 centimeters (7.9 inches) and is attached to the tractor by a three-point hitch.

No-till drill with fertilizer attachment made by Tye in USA. This model, Pasture Pleaser, is pulled by an integral hitch (1981).

No-till drill made by Tye uses three disks to cut through a dense crop residue (1981).

Model PL 2000 no-till corn planter with fertilizer attachment made by Semeato in Brazil. This model is attached to the tractor by a three-point hitch (1989).

No-till planter with fertilizer attachment made by Allis Chalmers in the USA. Model 300 is equipped with an air seeding unit. This was the first no-till equipment introduced into Chile in 1978.

Conservation tillage drill made by ICAT-INIA for planting wheat in minimum tilled or no-tilled fields that have small amounts of residue (Cauquenes County 1989).

however, the serrated disk does a better job because the design provides a better capability to cut the residue. Compared with the rippled and fluted coulters, the serrated disk can effectively cut an accurate depth with less movement of soil.

In conclusion, it is not necessary to till a wide strip of soil for good seed germination. In clayey soils the front disk must cut deeper than the double disk. This is particularly important during the first years under no-till because it helps get better root penetration.

No-till planters are required to cut residue and roots to do a good job. Because of this, the ease of adjusting the weight may be the big difference between planters. No-till planters need to have a strong, sturdy construction and be durable to support the added weight over each coulter.

Certain manufacturers have proven their planters can work without a front coulter. They have replaced it with offset double disk openers that work in the form of a V. With this system, one disk goes a little in front of the other, providing a better cut and penetration. The set of two double disks per planter unit permits one to locate the fertilizer in a furrow apart from the seed. This is a great advantage when using ammoniacal fertilizer while planting in humid areas with temperatures over 15°C (59°F) because if the seed is in contact with the fertilizer the toxic condition created by ammonia can damage the seed. This system requires only the two disks per planter unit for planting in the fall and winter. During this time of the year, the seed can be placed with the nitrogen fertilizer.

When the no-till planters are not equipped with a front coulter, but only with the double disk opener, the seed will be placed at the same depth as the bottom of the furrow. This makes it more difficult for the seedling roots to develop during the early stages of plant growth. This is especially true in the fine-textured soils, so it may be necessary to use a front coulter on these soils.

No-till planters also are manufactured with only one disk per planter unit. This disk cuts the crop residue roots at a depth sufficient to place the seed and fertilizer. The disk is large (45 centimeters or 18 inches) and follows at a 7 degree angle to the vertical. This angle permits the disk to cut the slit and at the same time firmly place the seed. This design is suitable for sandy soils that have weak structure.

In clayey soils that have good structure, the slit of the disk is of great importance for the seed to germinate and emerge from the slot area. Where other planting methods were used, the cotyledon had to look for a way to emerge. This caused slow germination and seedling emergence and in some cases the seedlings were lost. The double disk planter can be an advantage in these conditions. It also allows for the emergence of the plant to grow vertically and results in better seedling development.

Planters are available that place the seed at a controlled depth suited to the seed size and the microrelief of the ground. Generally, the more traditional planters are sold with depth bands to control the depth between 25 and 50 millimeters (1 and 2 inches). This accessory, if attached with scrapers, helps to place the seed and provide for uniform emergence of the plants.

A hopper that holds the fertilizer for application by banding during planting is needed because a significant part of the fertilizer must be placed next to the seed in the planting furrow. The application of all the phosphorus fertilizer next to the seedling root at the beginning of using the no-till planting allows for better availability of the nutrient compared with any method of traditional planting. In the case of other nutrients, the fertilizer can only be partially applied by the planter. If the availability of the fertilizer in the soil is low, the amount of nutrients applied with the seed can almost always satisfy the initial plant requirements. You also should consider the high solubility of the compounds that contain nitrogen and the consequences of their movement in the soil. Splitting the application of nitrogen during the period of vegetative growth of the plant assures the availability of this nutrient in the periods of critical and maximum demand.

Some planters are equipped with a third seed box, smaller than the others, for planting grass and legume seeds. This permits establishing pastures and forages without plowing. This seed box can be used for sowing canola and other small grains or for application of some granular pesticides.

The seed press wheels are used in loose, medium-textured (loamy sandy or silty) soils to produce a condition of good seed to soil contact, which favors germination. They are not needed for fine textured-soils.

Some no-till planters use an air chamber to measure and disperse large seeds such as corn and sunflowers. On some planters the air pressure is produced by a small electric fan run by the energy from the tractor battery. Other designs operate by a vacuum created by a moving fan driven by the PTO. In both cases, the mechanism to distribute the seeds uses the same principle.

The advantage offered by this type of planter is the uniform distribution of the seed. The system allows only one seed to be held in each cavity of the planting plate, which assures a placement of a certain number of seeds along the row. This type planter does not have seed tubes to conduct the flow of seeds to the furrow because the seed opening coulter is placed directly below the seed hopper. This permits the placement of the seed in the top few centimeters of the soil. Other equipment options include an electronic meter that indicates the number of seeds dropped by watching a monitor and an alarm that alerts the operator when there is an interruption in seed flow.

At Chequén, we use this type planter and can verify the precision that can be obtained with these corn planters. It is important to us to accurately monitor the high population of plants necessary to achieve our high yields.

The planter's plates are precise for the shape of the seed and must have seeds that are suitable for the openings. Various sized plates must be available, or the seed size must be uniform. This can be a disadvantage in the field situation because acquiring a large number of different plates to fit different seed sizes becomes objectionable. If the desired spacing and population is not achieved, the yield potential will suffer.

The market does not have many types of no-till drills. We can distinguish two types: (i) drills that attach to the three-point hydraulic hitch of the tractor, and (ii) drills that are pulled from the drawbar of the tractor.

The drills that attach to the three-point hydraulic hitch are lightly constructed. These drills are effective on flat soils, but require more horsepower than the pull type drills. The extra power is not for extra traction, but to lift the equipment. The pull type drills have hydraulic cylinders to lift or lower the machine. The body of the drill can operate independent of the tractor which allows for better planting movement on rolling topography.

Small grain drills for crops such as wheat, 15 to 25 centimeter (6 to 10 inch) row spacing, require about 5 horsepower, and large grain drills for crops such as corn and sunflowers at 61 to 81 centimeter (24 to 32 inch) row spacing, require 20 horsepower per row (if two-wheel drive). Tractors with four wheel drive require less horsepower per row and function faster, safer, and more economically. All equipment and tractors used in the management of no-till should have high flotation tires. The tires should have little or no tread, if possible, because the tread leaves tracks and compacts moist soil.

1.6.11. Regeneration of Pastures

Forage drills were brought into use in Chile in the mid 1960s. They were the first planters that really did not require prior soil tillage. Some of the models work on the principle of spinning

First no-till pasture seeder introduced in Chequén, made by Bush Hog in the USA (1964).

Main photo: Pasture seeder made by John Deere in the USA, Model 1550 attached to the three-point hitch of the tractor and driven by the power take off. Insert: detail of the cutting disk of the John Deere Model 1550 (Negrete County, 1990).

Rotary cultivator made by Dyna Drive in Spain (1990).

drill blades using the PT0. These blades till the soil in a narrow strip and then drop the seed and fertilizer behind.

The limitations in using forage drills include:

- Insufficient soil for covering the seed
- Limited soil for covering the seed caused by spinning blades that throw soil from the strip
- Accelerated erosion if drills are used in the direction of the slope
- High power requirement per row compared with the models that open the seed furrow with a disk or coulter
- Limited use on stony or rocky ground
- Higher hydraulic capacity [greater than 25 liters (6.6 gallons) per minute at a pressure of 2100 pounds] is required to adapt the drill to the three-point hitch of the tractor
- Poor results with crop residue on the soil surface

Some pasture drills are equipped with fluted coulters and/or small chisel points. All the forage drills have been pioneers in the introduction of no-till, but they fail to produce satisfactory results if the correct herbicide is not used and if experience in the use of a no-till system is lacking.

1.6.12. Rotary Cultivators

In this chapter, we have mentioned the microrelief as important in the establishment of no-till. Rotary cultivators work on the principal of turning blades or spokes (but not with the use of the PTO). This machinery is useful in clayey soils that require improvement in the microrelief.

The design is based on two parallel axles working together with distribution plates for the spokes. This equipment does a good job because the axles are united on one side by chain pulleys. The axle in the front is driven by a gear three times larger than that on the back. The front axle turns slower than that of the tractor, whereas the rear one moves faster than the front.

Rotary surface cultivators do a good job if operated at a proper speed so not to disperse the soil; therefore, minimally affecting the structure. In the process the spokes pull the weeds out by the roots, leaving them on the soil surface. By moving a small amount of soil, the equipment is adapted to the conservation tillage system. The equipment requires 40 horsepower per working meter (1 horsepower per working inch), tractor front wheel weights, four-wheel drive, and connects to a three-point hitch.

These units are manufactured in England, Spain, and Brazil. In Chequén, this equipment is used only for no-till in soils where the microrelief permits and to incorporate organic material in degraded soil. This is not a permanent alternative in the management of no-till.

BIBLIOGRAPHY

Aapresid. 1990. *Siembra directa.* Abril. Rosario, Argentina. p. 1.

Baer, E. 1986. Lupino: situación actual en Chile y sus perspectivas. *En IV Seminario Nacional de Leguminosas de Grano.* Universidad de Chile, Fac. Ciencias Agrícolas y Forestales. pp. 250–288.

Behn, E. 1977. *More profit with less tillage.* Wallace-Homested Book Co., Des Moines, Iowa. pp. 5–9.

Besoain, E. 1958. Mineralogía de las arcillas de algunos suelos volcánicos de Chile. *Agricultura Técnica* (Chile) 18(2):110–165.

Contreras, H. 1973. Curso de conservación de la naturaleza y sus recursos renovables por TV Nacional. U. de Chile. Fac. Ciencias Agrícolas y Forestales. *En* Clase 3: La erosión del suelo por acción de las aguas. 11 pp.

Delgado, L. 1983. Incidencia de la cobertura vegetal sobre las propiedades, desarrollo y clasificación de perfiles graníticos del área de Florida, VIII Región. U. de Concepción. Fac. Ciencias Agropecuarias y Forestales. Chillán, Chile. pp. 5x–9x. (Tesis de Grado).

Derpsch, R., and J. Alberini. 1982. Experience with cover crop lupins in the State of Parana, Brazil, and its importance for water erosion control. German Agency for Technical Cooperation, LTD. GTZ, Germany. IAPAR, Londrina, Brazil. pp. 193–194.

Endlicher, W. 1988. *Geoökologische undersuchugen zur landschaftsdergradation im küstenbergland von Conceptión* (Chile). Franz Steiner Verlag Weisbaden GMBH, Stuttgart. 192 pp.

Espinoza, W. 1968. Antecedentes bibliográficos del alofán y su determinación en los suelos de Ñuble. U. de Concepción, Escuela de Agronomía, Depto. Suelos. Circular Informativa No. 23. 48 pp.

Espinoza, W. 1973. Los suelos volcánicos chilenos: distribución, génesis y características. Universidad de Concepción, Escuela de Agronomía, Depto. Suelos. *Boletín Técnico de Suelos* No. 49. pp. 21–26.

Gischler, C. y C. Fernandez. 1984. Técnicas económicas para la conservación y gestión del agua en América Latina. UNESCO (ROSTLAC). Montevideo, Uruguay. pp. 11–17.

Lara, O., y O. Andrade. 1983. Una nueva enfermedad, mancha café (*Pleichaeta setosa* Kirchn. Hughes) en cultivos de lupino (*Lupinus albus* L.) en la IX Región. *Agricultura Técnica* (Chile) 43 (4):391–392.

Letelier, E. 1988. Efecto de las lluvias invernales sobre el manejo del cultivo de trigo en el secano costero de la VI Región de Chile. PROCISUR, Diálogo XXIV, Manejo y Conservación de suelos. IICA/BID. 78 pp.

Lowdermilk, W. 1953. Conquest of the land through 7,000 years. *Agriculture Information Bulletin* No. 99. USDA. pp. 1, 20–21.

Marelli, H. 1989. La erosión hídrica. INTA, Marcos Juárez, Argentina. (3):6–10.

Mironova, T.P., N.S. Kuptsov, and N.M. Pushnova. 1990. Feeding virtues of narrow-leaved lupin (*Lupinus angustifolius*). *In* 6th International Lupin Conference. International Lupin Association. Temuco, Pucón, Chile. p. 15.

Monsanto. 1988. Produciendo para el futuro: labranza conservacionista. Buenos Aires, Argentina. pp. 5–9.

Pendleton, J.W. 1979. Cropping practices. *In* Maize. Ciba-Geigy Agrochemicals. Switzerland. pp. 18–21.

Peña, L. 1986. Control de la erosión hídrica en suelos volcánicos Dystrandept, mediante labranza de conservación. *Agro-Ciencia* (Chile) 1(2):59–64.

Raggi, R.M. 1989. La agricultura de riego en Chile. Situación actual del riego, del dranaje y perspectivas. *En* Taller Técino, Examen de mecanismos de degradación y de metodologías en el manejo de aguas y suelos de tierras bajo riego. Mendoza, Argentina. FAO/GCP/RLA/084/JPN. pp. 105–128.

Riquelme, J., B. Fernandez, y L. Peña. 1972. Influencia de algunas prácticas de manejo en viñas plantadas en lomajes de la costa, sobre el proceso de erosión hídrica. U. de Concepción, Escuela de Agronomía. Depto. Suelos. *Boletín Técnico de Suelos* No. 40.

Schenkel, G., y H. Bleiholder. 1967. Factores que afectan a la determinación de la fijación de fósforo en algunos suelos chilenos. U. de Concepción, Escuela de Agronomía, Depto. Suelos. Circular Informativa No. 15. 27 pp.

Soza, R. 1980. La cero labranza en el cultivo del maíz. *Tecnología y Agricultura* (Chile) 2(9):17–24.

Yoo, K.H., and J.T. Touchton. 1989. Runoff and soil loss by crop growth stage under three cotton tillage systems. *J. Soil and Water Conservation* 44(3):225–258.

CHAPTER 2

The Management of No-Till

2.1. Sloping Soils and Their Management

The most serious limitation of arable land is slope because of the high risk of soil loss by erosion. The greater the slope, the greater can be the loss. The Coastal Range soils that make up most of Chequén, as well as the Andean foothill soils, are on rough topography that is moderately to steeply sloping.

With the advent of no-till, the topographic aspects have been left in second place. Today, one can say that with **no-till, the slope is not a limitation in soil conservation, but more than anything a problem relative to efficiency and safety in driving equipment and machinery**.

No-till, a technique that does not disturb the soil, ensures the physical stability of the soil, allowing its use on steep slopes; however, to work soils with steep slopes complicates the management of the system. The main problem observed is slower speed of the work. This results because of the reduction in the velocity of the machinery, especially when seeding. In this circumstance, due to the lateral inclination of the drill, velocities above 4.0 kilometers per hour (2.5 miles per hour) tend to lift seeds out of the furrow because of the rubbing on the inner walls of the drill disks. For this reason it is important to slow down equipment operating speed.

Steep lands can be farmed with a no-till system as long as the equipment is properly managed. Under these conditions, which exist on Chequén, the tractors should have four-wheel drive and the implements should be pulled by the drawbar, especially the drills. Pulling the equipment with the drawbar is safer and more efficient on steep lands. Gentle slopes offer unique benefits for no-till because the excess rainfall can run off, avoiding ponding water on the planted crop. This is particularly important in high rainfall areas in which prolonged moisture excesses can affect the crops.

Tractor-drawn combine, on steep slopes, with an operator-activated hydraulic system that allows the combine to be leveled and, thereby, reduces grain loss (1989).

By not disturbing the soil, no-till allows seeding in the direction of the slope. This has the advantage of a more uniform distribution and covering of the seed and better use of the area to be seeded. In these cases the soil should have adequate mulch, not have excess moisture, and have a structure that guarantees traction of the tractor on the soil. Four- wheel drive tractors should have counterweights in the front.

Seeding on the contour or across the slope can leave seeds uncovered because lateral inclination of the drill would make the disks at the upper end go shallower, which is common when the slope exceeds 10%. Also, under these seeding conditions, the furrow can remain uncovered because the disk on the down-slope end removes soil, leaving the seed bare. In these cases, it is ideal to work in the direction of the slope to avoid the lateral inclination of the drill.

The use of no-till in sloping areas allows for expanding crop areas and, according to what was tried on Chequén, producing grain where before it could not be done. This fact has been very important in farming formerly marginal soils. Proper management of the crop residue is indispensable to complement efficient conservation; and the physical stability of the soil depends, in large measure, on this.

2.2. MICRORELIEF

One of the principal obstacles that can occur in the establishment of no-till compared with a cultivated soil is microrelief. In a cultivated soil, the drill disks enter with ease and the seed is left in close contact with the soil and at the proper depth. The soil to be no-till seeded is not tilled; therefore, the best conditions needed to place the seed at a correct and uniform depth must be assured. Strong microrelief can seriously affect the seeding by low seed density and uneven germination.

Microrelief can be defined as small topographic alterations of the soil surface that can vary from 2 to more than 10 centimeters (1 to 4 inches). The smallest alterations can be caused by the trampling of animals. Prior tillage implements (plows and disks) and the inappropriate traffic of drawn agricultural machinery and wagons on wet soil cause major alterations.

Erosion also can change the microrelief of the soil, leaving channels that can affect the passing of drills or other equipment. In these cases, a leveling blade is preferable to fill the rills or channels made by the water in previous crops; or a shovel can be used. To prevent erosion where the soil is removed, 2 to 3 tonnes per hectare (1800 to 2700 pounds per acre) of organic residue should be put on the bare surface.

The soils most affected by an uneven surface are those with fine texture. This is not true for the trumaos soils that are predominantly medium-textured. In the fine-textured soils, this uneven condition can become serious in no-till because of poor plant emergence. This would necessitate superficially disking the soil and later passing over it with a leveling drag. If the coarser textured soils including the trumaos do not have straw or stubble from previous crops, they only need to be passed over with a leveling drag. A soil with slight microrelief can be seeded without disking or leveling as long as the seeding is done at low velocities (3 to 4 kilometers per hour; 2 to 2.5 miles per hour). Natural leveling will occur over time if animals or machinery are kept off wet soils.

Tillage tools, especially plows and disks, can leave deep furrows and/or alter the degree of microtopography. This condition is intensified when the tools are used improperly. Because of this, a previous conditioning of the soil is required before no-till seeding.

Serious management problems can occur by no-till seeding in areas with severe microrelief. This situation eventually necessitates smoothing the soil surface. As well as improving the popu-

Trampling by cattle on moist soils can be a serious limitation in the management of no-till (1991).

lation density of the crop, soil with a uniform surface permits increasing the velocity of all the tasks of the seedings. This brings about greater efficiency and yields along with lower costs of operation.

No-till seeding is ideal on a pasture as long as a uniform surface condition exists. Although livestock use the forage, it is not suitable to include pasture in the crop rotation. The no-till system already permits the use of the soil in the most intense manner. The fact that the soil is not disturbed, together with the management given to the stubble, generates and improves the conditions much like a pasture can produce when it is included in a traditionally used crop rotation. Pasture is not included in the no-till crop rotation to avoid the compaction and the microrelief typically occurring in the management of livestock.

2.3. Land Use Capability and Classification

At one time, the USA system of soil capability classification was used on Chequén. This classification system limits areas under tillage mainly by slope and separates the soils basically into two large groups, arable and nonarable.

At Chequén, this soil classification into Land Use Capability Classes was an ordering of the soils that indicated their relative adaptability to crops and the difficulty and risks that were present in using them. This arrangement, still used to some extent in the USA, is based on the capacity of the land to produce and indicates the natural limitations of the soils in relation to the agricultural systems that are used.

The first group is made up of arable soils distinguished into Capability Classes I to IV. The group of nonarable soils are classified into Capability Classes V to VIII.

Capability Class I soils have very few limitations that restrict their use. That is, they are suitable for any adapted crop of the area. They are deep, nearly level, well drained soils that are easy to work. Their capacity to retain moisture and their natural fertility are good. They require simple management practices to maintain their productivity and conserve their natural fertility.

Capability Class II soils have slight limitations, especially with reference to fertility, depth, drainage, and slope. These limitations can occur alone or can be combined in such a way that reduces the selection of crops or requires moderate conservation practices.

Capability Class III soils have moderate limitations that restrict the selection of crops. The slope can vary from level to moderately sloping, an aspect that severely complicates gravity irrigation. These soils require more careful conservation and management practices.

Capability Class IV soils have severe limitations that significantly restrict the selection of the crop. These soils require careful and complete conservation and management practices that are more difficult to apply and maintain than the soils of previous classes.

Capability Class V is special because it can include level or nearly level, stony, flooded, wet, soils suitable only for wildlife. They are normally found in old riverbeds. In general, this capability class is uncommon in traditional agricultural areas.

Capability Class VI soils are unsuitable for the traditional tillage methods, and their use is restricted to pasture and woodland. The soils of this class have permanent limitations that normally cannot be corrected. The limitations include very steep slope, susceptibility to severe erosion, excessive stoniness, shallow root zone, excessive wetness or doughtiness, and high content of salts.

Capability Class VII soils have very severe limitations that make them unsuitable for crops mainly because of their excessive slope. Their principal use is woodland and permanent pasture.

Finally, Capability Class VIII soils correspond to soils without agricultural, grazing, or woodland value. Their use is destined only for wildlife, recreation, and watershed protection. These soils are common in intermediate and high areas of the Andes Mountains.

The Land Use Capability System was created by H.H. Bennett and is still promoted today by the U.S. Department of Agriculture, Natural Resources Conservation Service. This classification system was useful on Chequén as long as we didn't have a better system of conservation and production.

No-till has changed this because we can now grow crops in areas where previously the high risk of equipment operation and accelerated loss of soil fertility did not allow it. All this makes us think that the normal grouping of agricultural soils into arable and nonarable is obsolete because it is demonstrated that cash crops can be produced economically in some soils grouped as nonarable without any risk of erosion. This is possible because no-till does not disturb the soil.

2.4. STUBBLE MANAGEMENT

A productive farm with high grain yields also generally produces high quantities of crop residue. The management of this residue in no-till has been converted into a true art, which has direct relation with the success in soil conservation and a profitable farming operation.

In Chile and in many other countries, the farmers' rejection of their own crop residue and stubble is a historic phenomenon. They try burning it, which sometimes damages their own property or leads to legal problems for damage caused to their neighbors property.

All these risks and damages allow them, without any technical justification, to leave their soils as clean as possible. This can be one of the more degrading practices in soil management and is responsible for the loss of fertility.

The traditional tillage systems generally require the removal of residue because any excess that is not well managed can cause a problem while cultivating it. Operations with traditional tools are difficult because of the obstruction, or when the residue piles up due to plugging. This can affect the yield of the next crop. The physical problems created by badly managed residue provoke farmers to burn it or move it out of the field before planting. We also observe that in spring crops, and after the last harrow, even the loose roots are taken out of the field. The stubble and this part of the roots are very important to soil fertility and structure. **To burn or take out this valuable organic material is an attempt against the life of the soil.**

The slow loss of organic matter results when the soil is deprived of the residue produced by the crop. This means a slow loss of fertility and affects soil productivity. Soils that have a low level of organic matter (less than 2%) depend more and more on chemical fertilizers, especially nitrogen. It is more difficult for the soil to increase its natural organic matter than to resist losing it. According to research conducted on Chequén, the only practical way to achieve greater soil organic matter is keeping **the stubble on the soil surface and then planting the field without disturbing the surface.**

Adequate equipment is available to efficiently manage the crop residue, such as the choppers and mowers mentioned in Chapter 1. As more stubble is left on the soil, it should be chopped more to reduce its volume and to facilitate the planting. On some occasions I have observed that the stubble has been chopped and then partially incorporated into the soil (performing what we know as minimum tillage). This is somewhat acceptable in soil conservation; however, this way of management involves the moving or disturbance of the soil, leaving it susceptible to erosion. In addition, the nutrient availability to the crop to be established varies because of incorporation of residue into the mineral profile.

Turning under or mixing crop residue into the soil produces an excess of carbon dioxide, methyl alcohol, and weak acids that can interfere with the physiological process of soil metabolism. In this case the anaerobic activity is much greater than aerobic. The residue decomposition in the soil must happen as slowly as possible to keep humus material from its final degradation or mineral form. We have found that humus produced naturally by farm residue and left on the soil is the main reason for the spectacular fertility increase in the soils of Chequén. Plowing or disking residue into the soil speeds up soil organic matter decomposition, and less humus and organic matter are in the soil profile. For these reasons conservation or minimum tillage has not been established at Chequén. We always plant directly into the crop residue.

In relation to this, I think it is interesting to point out my experiences while making presentations in Mexico, Brazil, Bolivia, Uruguay, Argentina, USA, England, and in my own country. It has been profoundly called to my attention by farm professionals that although they cautiously accept the no-till system, they insist that the crop residue should be incorporated into the soil to accelerate its decomposition. I think that, from a theoretical point of view, it is hard to imagine that the natural process of organic material incorporation and its humus formation and mineralization inside the mineral profile could be faster than the process that happens over the soil surface with adequate moisture and temperature conditions. According to Chequén research, these processes are faster than they are normally supposed to be. Therefore, it seems that the only way to completely understand this situation is through direct observation.

2.4.1. Alternatives in Stubble Use

The strongly industrialized world in which we live requires larger quantities of natural fiber every day. This fiber has a wide variety of farming and industrial uses.

2.4.1.1. Agricultural Uses

Proper management is required to plant into small grain stubble because it is not always possible to leave the entire amount over the soil. The Friends of the Soil Association, from the Argentina Republic, highlight the importance of crop residue in livestock feeding. A 65% assimilation of crop residue can constitute an excellent food ration if it is properly complemented with other food nutrients. Beef cattle, especially the female, consumes cereal straw in the season when green pasture is scarce. Another interesting aspect is that the nonassimilable fraction, that is 35% of the consumed straw, is converted to manure by the cattle. This semi-decomposed product benefits the soil and the crops to be planted (Friends of the Soil Association, 1987).

Many formulas are available to calculate the crop residue surplus to use as livestock feed; however, the straw must be previously baled so it can be treated with liquid ammonia, sodium hydroxide, or another procedure. These treatments are effective, but require careful management of these chemical compounds. The cattle get a better assimilation of the products rich in fiber with these treatments.

At Chequén, the surplus of small cereal residue (wheat, oats, triticale, barley, and rye) has been used directly from the field. The beef cattle (Hereford) efficiently graze and digest the stubble. One observed problem is that the straw, no matter if it is chopped, has low palatability. The following procedures can improve its use:

- Chop the straw with a chopper
- Mix 100 kilograms (221 pounds) of molasses, 5 kilograms (11 pounds) of urea, and 100 liters (26 gallons) of water in a storage tank. This mixture yields 205 kilograms (452 pounds) and is enough to treat 1 tonne (2205 pounds) of straw.

Excess stubble is feed for cattle (1986).

Knowing the amount of residue per hectare, you can calculate the required mixture needed. To apply this mixture evenly and directly over the stubble, we use a honey wagon with two extendible pipelines in the back and two partial angle spreaders applying 20 liters per minute (5.3 gallons) each 8 meters (26 feet) apart at the end of each pipeline. The pressure (30 pounds) was achieved with a high pressure flow pump using the tractor power take off (Crovetto, 1987). The molasses mixture improves the straw palatability, and the urea improves the protein value, which results in a higher forage ingestion per animal and a significant weight gain.

The field yield must be known to calculate the stubble amount per hectare. For cereals, the residue yield rate is about 1:1, so for every 1 tonne of grain produced there is 1 ton of stubble; however, the best way to calculate the stubble is to clip 1 square meter before threshing at three or four random sites, take out the grain, and then calculate the average. Multiplying this amount by 10,000 gives the stubble amount per hectare. I believe this is a more precise and practical way to calculate the yield, especially for farmers being initiated in no-till. The participation will commit them to the system in a way that gets the needed observation discipline required to become a good no-till farmer.

The cattle must not consume all the stubble. A minimum of 3 tonnes per hectare (2700 pounds per acre) must be left over the soil.

If the field with stubble will not be planted soon, it is not necessary to remove the stubble or to apply the molasses mixture, but the stubble should be chopped and evenly distributed over the field. As explained in Section 2.5., corn, canola, and grain legume residue do not present allelopathic problems in future field management using the no-till system.

A molasses mixture dissolved in water and sprayed on the stubble is a good alternative for improving the palatability of residue (1988).

Windrowing and baling the straw or hay reduces the allelopathic effects in the crop that follows (1991).

When cattle are not available to use the surplus stubble, the residue can be managed in the following ways:

- Leave the chopped residues. Windrow the straw with a side delivery rake leaving a 8 to 16 meter (25 to 50 feet) wide strip, clear of straw where the next crop will be planted. The straw placed in windrows should be contained in 1.0 or 1.5 meter (3 to 5 feet) wide strips. These straw windrows cannot be planted immediately. They need time to reduce the straw volume and the allelopathic effects. The surface area to be planted is reduced by the area occupied by the windrows. In the next year the straw in the rows could be returned to the original place using a rotary chopper. With these conditions, and after a year of being piled, the straw will be almost decomposed and will constitute an excellent way to save the soil moisture, thereby achieving higher yields in the next crop to be planted.
- Bale the windrowed straw left by the harvester.
- Take off the crop residue leaving 3 tonnes per hectare (2700 pounds per acre) using a forage chopper to make piles on the edge of the field or store them in the barn.
- Add straw during the ensilage process of high moisture forage so that nutrients lost by seepage can be absorbed by the straw.
- Use the straw as mulch in eroded areas.
- Use it as a winter floor in the barns, and recycle it later as a cover for meadows or other eroded areas with the additional benefit of the nutrients from the livestock wastes.
- Use it as food in worm production mixing it with manure and other residues.
- Prepare compost with the straw.

Since 1992, the small grain straw at Chequén Farm has been raked into piles 1.5 meters (5 feet) wide and 10 meters (33 feet) apart. Less than 1 tonne per hectare of straw was left on the sur-

Straw on the soil surface is harvested into square bales and stockpiled. This is not a good alternative for improving soil quality (East Auglia, England, 1993).

face. This process would work best on large fields. It is fast and easy to do, costs are low, and effective final results are produced. Most importantly, the residue is kept on the field. Residues that are not allelopathic do not need to be piled.

2.4.1.2. Industrial Uses

The problem of using the stubble outside the field is generally directly related to the cost factor, mainly the high cost of transportation. If the product can be compacted mechanically and transportation costs can be lowered, then marketing will be possible. The pelleted or baled straw has a wide market including livestock feeding, packing filler material, insulation material, raw material for paper or cellulose manufacture, and also as fuel.

2.5. ALLELOPATHY

2.5.1. General

Allelopathy occurs at the beginning of the decomposition of crop residue. It can affect some seed germination seriously and also can inhibit seedling development.

Allelopathy (a synonym for phytotoxicity) can be defined as a natural biological-chemical interaction in which a plant or its residue damages or interferes in the development of other plants. These substances, called allelochemicals, also can damage seeds and seedlings from the same crop.

In 1937, the Austrian botanist Hans Molish was the first scientist to observe this phenomenon. He defined it as toxins generated by microorganisms, such as bacteria, actinomycetes, fungus, and algae. In this chemical interaction, species that exude phytotoxic compounds damage other living things that are sensitive to the effect.

2.5.2. Allelopathy Effect in No-Till

Behmer and McCalla (quoted by Elliott et al., 1978) identified the "patulina," originated by the fungus *Penicillium urticae,* as a powerful inhibitor agent of wheat planted into stubble from other small grains.

Stowe, mentioned by Elliott et al. (1978), indicates that other organic components, such as hormones and enzymes, when produced in excess, can inhibit development of seeds and plants. In 1983, Verónica Campos, botanist from the University of Concepción, found allelochemicals and their phytotoxic properties during an oat residue research project. She found the presence of ferulic acid and *p*-cumaric, important compounds generated during residue decomposition in no-till management systems.

These allelochemicals act directly on the plant, inhibiting the photosynthesis and enzymatic activities and also growth regulators, thus changing the plant's protein synthesis. The ferulic acid inhibits some physiologic activities, such as potassium, calcium, manganese, and iron ion permeability, at the radical site in oats (*Avena sativa* L.; Campos, 1983).

Dr. Francisco Pérez, scientist from the University of Chile, found low molecular weight chemical compounds exuded through the wild oat roots (*A. fatua* L.) as escopoletina and vanillic acid, which are phytotoxic to the wheat plant. A thick growth of wild oat root can reduce the wheat yield by its allelopathic effects (Pérez, 1990).

Allelopathy is basically generated during the first steps of organic material decomposition, which coincide with the autumn rainfalls. Its intense effect from crop residue in the soil damages seeds in the germination stage and the young seedlings already established. This directly affects no-till crops, and it can damage reduced tillage crops where the stubble is somewhat incorporated and the soil has not received enough rain or the soil moisture has been deficient for more than 15 days. Under normal moisture conditions, clayey and moist, humid soils can deactivate these organic compounds, causing less allelopathic damage to the recently planted crops.

Although some literature mentions allelopathy produced by the decomposition of green tissue, at Chequén we have not observed damages to corn, lupin, or sunflower crops after weed control with herbicides was performed in the winter–spring crop. We have observed that summer crops benefit from the winter–spring cover crop because of improved soil fertility.

Crops planted in well managed wheat stubble do not produce negative allelopathy effects. This may indicate that during this agroclimatic period some of the microorganism and fungus that produce allelopathic compounds do not have an adequate medium for their proliferation.

2.5.3. Rye, Barley, Wheat, Triticale, and Oat Stubble

Moisture is needed for the decomposition of organic residue. Crops planted in the stubble of small grains, such as rye, barley, wheat, triticale, and oats, have a high risk for allelochemicals production when the stubble has sufficient moisture. Some microorganisms responsible for the initial stubble decomposition are the same ones that produce allelochemicals, those organic acids that have allelopathic effect. These can inhibit or destroy a crop if the crop is planted during the first 6 weeks of the rainy season.

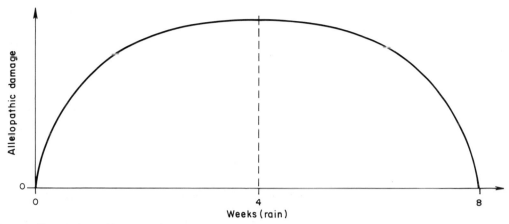

Fig. 10. Effect of allelopathic damage in relation to climatological conditions and time.

Guenzi mentioned that the allelochemicals act in a progressively stronger way until the fourth week and in a decreasing way until the eighth week, a situation that coincides with the beginning of the autumn rainfalls (Elliott et al., 1978), shown in Fig. 10. This chronological situation also is dependent on the quantity and type of residue that is covering the land. If the residue quantity exceeds 3 tonnes per hectare (2700 pounds per acre), more time is needed to avoid negative allelopathic effects.

For the allelopathic effect to happen, the residue needs to be in contact with and forming a dense cover over the soil. The risk of damage for the crop is low if the planting is performed in upright stubble and not chopped. This system is used in the USA Corn Belt, where snow covers the soil surface during the winter and there is a special interest to preserve it until the next spring.

The situation is somewhat different in Chile because the chopped stubble on the soil is an effective way to conserve soil and water during winters without snow. Another aspect to be considered is that large amounts of wheat stubble are produced in this country, and that it is important to make windrows with the stubble left by the combine. If the straw surplus is not removed, the alternatives would be to perform the planting at the end of winter or the beginning of spring, or to plant the next year. The alternatives mentioned in Section 2.4.1. also could be used. Farmers oppose this because eventually they will lose one crop year. I believe it is important to point out the benefits of not planting one year. A good alternative would be to broadcast a mixture of oats and vetch with minimum fertilization. These forage species germinate in the stubble and provide winter forage to the cattle. This forage can be cut and removed from the field with a forage chopper, made into silage, or baled in the spring.

If care is taken to apply a systemic or contact herbicide after the cutting and before the weeds produce seeds, a good weed control program is achieved and a successful crop is ensured for the next year. All of this is accomplished without burning or taking the crop residue out of the field.

According to research conducted at Chequén, no phytotoxic effect occurs if the wheat crops are planted at the end of winter (August) with quantities of stubble lower than 3 tonnes per hectare (2700 pounds per acre); however, it is appropriate to reiterate that the stubble should be chopped, well distributed, and left over the soil exposed to the rains for at least 6 weeks.

Late winter planting in areas with high quantities of stubble over the soil can be an advantage because of the higher moisture availability for the winter–spring crops. In continuous rotations, the full use of spring crop residue also will improve the soil's fertility and productivity.

Allelopathic effects in small grain straw. The chopped straw is infected by fungal action and the production of allelopathic chemicals that can destroy seeds and seedlings (1986).

When wheat is seeded in corn stubble, no allelopathic effects have been found (1989).

In no-till, the weed population has a tendency to change (Soza, 1980). The narrowleaf plants (monocotyledons) have a tendency to increase and the broadleaf plants (dicotyledons) to decrease. In areas without a crop rotation, we have observed a strong invasion of dog fennel (*Anthemis cotula* L.) in dry land crops. This is an insignificant problem in crops in rotation with corn, sunflowers, or other crops. This can indicate that some weeds are not sensitive to allelochemicals.

2.5.4. Corn Stubble

Corn for grain is an important production factor at Chequén. Besides its high yields and many benefits to the production in the field, it also contributes a large benefit to the commercial poultry business. The high quantities of residue produced, 12 to 20 tonnes per hectare (11,000 to 18,000 pounds per acre), provide adequate weed control for the next fall crop. Excellent weed control is obtained in wheat, oats, lupin, and other crops without applying a preplanting herbicide following a corn crop.

Corn is planted every 2 years in the same field establishing a rotation with wheat and lupin. The wheat stubble has a positive allelopathic action that helps control broadleaf weeds. Some green vegetative cover also helps keep the soil biologically active during the cold season. Because of better weed control in the last few years, oats or rye (only as green winter cover) has been established during the fall and with wheat stubble in rotation with corn. This improves the cover and has the same advantages mentioned above. Kimber (1967) studied toxin production in several grains and legume stubble. He determined that the extracts are more phytotoxic in green stages than in mature or dry residue. Our experience, however, is that the green cover from fall-spring

Chopping of corn stubble is important to obtain a uniform fall seeding (1989).

crops, especially those treated with systemic herbicides, has not shown allelopathic effects on germinating corn seeds nor in young corn plants.

We have not seen allelopathic effects when wheat and other cereal crops are planted into corn, sunflower, lupin, or canola residue. This means that there are no limitations for crops planted into these types of stubble.

Rye residue has major allelopathic effects against weeds. In the USA Corn Belt, rye is commonly planted by aerial seeding before harvesting corn. Besides keeping the soil biologically active during the cold winter months, a notable reduction of annual broadleaf weeds has been observed. This reduction is because of the allelopathic effect caused by the rye after it is cut and left over the soil 40 days before the corn is planted. If the rye is not more than 40 centimeters (16 inches) high and is not cut, then it is necessary to apply the crops normal preemergence herbicide.

The weeds in the thick stubble layer left over the soil are in various germination stages and some never become established. Grass weeds, such as annual ryegrass (*Lolium multiflorium* Lam.), is a strong competitor that can require chemical control at certain times. Seeds of summer weeds, such as crabgrass [*Digitaria sanguinalis* (L.) Scop.] and barnyard grass [*Echinochloa crusgalli* (L.) P. Beauv.], do not germinate in cold seasons, so there is no allelopathic effect on them. With time, these annual grass weeds cover more and more field area and it becomes necessary to control them with herbicides. This control is easy if, instead of corn, sunflowers or another broadleaf crop is planted. This rotation permits the application of a postemergence grass weed herbicide.

The germinated plants affected by allelopathy show chlorotic symptoms, such as light color, thin stem, weak and shallow roots, and a general lack of vigor. This can be confused with a lack of nitrogen, which also commonly happens in no-till (Crovetto, 1987).

Chopped lupin stubble has no allelopathic effect on wheat seedlings.

2.5.5. Lupin Stubble

The stubble of this legume does not produce allelopathic effects, so any grain can be planted into the residue. Because the lupin stem is somewhat bushy, its residue is easy and convenient to chop and can be spread very thin over the soil surface.

2.5.6. Rape or Canola Stubble

Canola is an important crop in this country. At Chequén, we have had good experiences with this crop and its residue. According to observation from the Montpelier Farm in Region VIII, no-till canola shows that its crop residue does not have an allelopathic effect. Most of the farmers that have no-till canola burn their stubble, so not much is known regarding these effects. Burning the canola stubble is not justified because it is an excellent base to plant any grain with no-till and does not cause allelopathic effects for the next crops. Canola stubble should be chopped and spread before the needed herbicides are applied for the next crop.

2.5.7. Sunflower Stubble

At Chequén, the sunflower crop has been researched for only 2 years, so there is not enough experience to judge its allelopathic effect in no-till. We can say that this residue should not produce a negative allelopathic effect in fall or winter grain crops as long as it is well managed.

In summary, allelopathy is a phenomenon originated under certain conditions that can have positive or negative effects on the planted crop or in weed control. The experience at Chequén shows that the stubble, especially that of small grains, needs to be managed carefully for fall planting. It is important to remember that the crop rotation can be helpful in managing allelopathic effects. Our experience indicates that wheat and rye straw produce a strong allelopathic action if the stubble is more than 3 tonnes per hectare (2700 pounds per acre). According to this and what the research people have concluded, small grain stubble can have a strong allelopathic effect, with rye stubble being the most active.

The problems caused by allelopathy can be avoided. The best method is not to exceed 3 tonnes per hectare (2700 pounds per acre) of small grain stubble and to plant after the sixth week following the beginning of the fall rains. We insist on efficient stubble management. At Chequén, we had good results applying 1 tonne per hectare (900 pounds per acre) of lime over the stubble to rectify the brief acid pH produced by the allelochemicals. Doses from 100 to 200 kilograms per hectare (90 to 180 pounds per acre) of lime can be enough if they are applied with the seed. In soils with pH lower than 6.0, the quantity of lime should be increased.

2.6. THE BURNING

Professor and Historian Augusto Vivaldi (1990) recounts an article from *El Ferrocarril* newspaper in 1870 regarding fire and its destruction of natural resources.

"It is not like in the foundry or the axe of the woodsman carrying load by load of wood to the furnace. It is the torch of a forest arsonist, which penetrates in the densest area of the mountain and after having formed a bed of dried twigs, half a yardstick high, at the foot of the secular trees, sets a fire in 20 or 30 places at once, converting a splendid forest not into a bonfire, but into a general fire in the extension of 50 or 100 quarter sections. A truly druidic forest, which amazes the Europeans and every cultural man, presents the sight of dead trees of which the most colossal ones still keep themselves standing.

The preservation and wise use of stubble helps to reduce pollution, slow down the erosive processes, improve soil fertility, and beautify the environment (1988).

The land covered by these burnt trees is an ash layer that the caretaker, having no need to work on,
throws some bushels of wheat to obtain a regular harvest for only two or three years.

Chile is being fired on behalf of agriculture, the southern desert is being decreed on behalf of the
wheat, likewise the northern desert was decreed and consumed on behalf of copper. We are not working, but enjoying the country. We are wasting our resources now, which will make work impossible in
future years. We are asking not for the product of the year, but for the capital too, to impoverish the
coming years."

This story is so heartless that it seems it is not related to Chile. It is difficult to believe that
we still use fire after being warned of its consequences 125 years ago. How is it possible that after
so many generations we do not take seriously the warnings and are not capable of recognizing how
fire has been destroying our country? In some way those responsible for the fires and its consequences are the uneducated farmers that cannot understand the results of their actions; however,
other farmers and foresters also use burning, knowing its effect and consequences. Regarding this
we ask: Why do our politicians, scientists, journalists, educators, and citizens accept these unfor-
tunate actions with nature that affect the natural biotic equilibriums and human life?

2.6.1. Burning and the Stubble

Chile has excellent climate conditions for grain production. This is especially true in the dry
coastal areas and next to the Andean Mountains. A cold and rainy winter and a moderate spring
with sunny days are adequate environmental conditions for good production of cereal crops.

The burning of stubble is a common practice in Chile. It damages the soil and reduces its fertility (1989).

The City of Temuco is polluted by smoke produced by the burning of stubble in the region (1989).

The calcination of woody residue can change the soil's physical, chemical, and biological characteristics. Calcination is the burning of organic material with the production of ash. Much of the material may be lost by gasification. Observe the soil's reddish color caused by fire next to the burned trunk (Florida County, 1981).

Genetics are advancing in creating new high yielding cereal varieties with less residue production in relation to the yield. Generally large amounts of residue are produced, but on many occasions, farmers because of their ignorance, reduce crop residue with fire.

Organic matter is the most significant factor in the conservation and improvement of soil productivity properties. It creates life and builds life through time. It is basically in two forms: one life represented by the animal and plant kingdom and the other by plants and animals that have ended their cycle and now contribute to the formation of humus. The last are fundamental in the creation of new life. With these examples, I am trying to explain that it is difficult to maintain an active media that lacks organic residue.

Dr. Hugo Zunino (1982) said that the organic matter is a decisive compound in the conservation of soil fertility. Conservation and improvement of soils requires careful management that can only be achieved by scientific knowledge of its composition. The relation between the inorganic and biotic components and their interrelation with the physical and chemical aspects of the soil must be understood.

Crop residue, along with the soil, provides the farmer with a great opportunity to strengthen the biotic aspect of the farm and make it more productive in the future. This important factor restores to the soil a high percentage of the nutrient elements extracted by the crops.

Antagonism does exist between residue and burning. The residue constitutes a base of the transformation into new life; the burning is responsible for its death.

2.6.2. The Fire and the Soil

Burning, a very old practice in farming, is an irresponsible way to reduce crop residues. Farmers using fire to reduce residue are unaware of the results of burning. They only think of immediate economic reasons and believe that the mineral salt left as residue from the stubble burning is important when it is incorporated into the soil during tillage. All this happens without considering the negative effects in the long and short run. The soil's natural parameters are affected, especially the biologic environment and the productivity.

The soils are exhausted in proportion to the frequency and intensity of the burning. The lack of soil fertility in Chile is basically caused by the extreme loss of organic matter, crop residue extraction because of calcination, and by erosion. It is estimated that in 1988, 400,000 hectares (988,000 acres) were burned, destroying 1,200,000 tonnes (1,323,000 tons) of crop residue. This is equivalent to about 25% of the surface area planted in the country. Burning is the least ecological and most degrading way to prepare a soil to be planted.

2.6.3. Physical Aspects

Nature, through time, has been capable of building productive soils over the base of a series of natural factors, such as climate, parent material, topography, and vegetation type. Without a doubt, the presence of vegetation and animals has accelerated the development of the weathering process on the Earth's surface resulting in the formation of new fertile and productive organic soils.

The lack of organic matter leaves the soil susceptible to erosion, reduces its productivity, increases costs, and destroys the landscape.

Generally, the volcanic silty soils (trumaos) are tilled unnecessarily after being stripped of their stubble by fire (Vilcun County, 1990).

Each year, the fallen leaves from grass, shrubs, and trees add a new, thin organic cover over the soil. This cover protects the soil from erosive processes and accelerates the pedogenetic process of soil formation. **Nature has shown that there should always be some organic residue in the soil's surface.** This is the fundamental base for the soil's productivity and conservation, and even for life!

When the soils are bare as a result of burning or tillage, they are exposed to water erosion. The rain has a demolishing effect over the bare soil. Each raindrop falls with an average velocity of 3 meters per second (10 feet per second) depending upon the size of the drop. The powerful gravitational energy generated when the raindrop hits the bare soil is the main cause of water erosion. This erosion affects the flat and the steep soils. The transportation of soil particles, however, is strongly influenced by the slope steepness; the steeper the slope, the higher the soil loss by erosion (see Section 1.2.4. and Fig. 4).

The first physical action of the residue over the soil is to dissipate the kinetic energy of the raindrop. It lets the water slowly contact the mineral soil, which absorbs it slowly and gradually. Thus, more water is infiltrated than runs off. This is the first step toward conservation.

The stubble on the soil is especially important in forming organic colloids that can improve soil structure and stability. The decomposition of the cellulose that occurs under natural environmental conditions from the residue is caused by some bacteria from the genus *Cytophaga* and *Pseudomonas*. This results in an abundant production of organic colloids. These colloids are attributed to the higher resistance of the soil to the raindrop impact (Molina, 1949). Polyuronic colloidal compounds help stabilize soil aggregates and protect soil against the impact of raindrops.

Water erosion is different in flatter, bare soils because although the soil loss is minimal, the soil can become compacted. The surface becomes crusted, which prevents normal rainfall infiltration. This reduces the storage volume and the plant population because of germination diffi-culties. In these circumstances, the soil will be completely dry within a few days even after intense rainfall. This is the most critical problem most countries suffer because it results in reduced food production.

In addition to the great benefits obtained when the crop residue is maintained over the sur-face and when the soil is no-tilled, a true microclimate is created. This microclimate protects the soil from high and low temperatures, reduces water evaporation or loss, and improves the physi-ological conditions for crop development (see Chapter 3).

If stubble burning in clayey soils is too intense, the primary chemical compounds can be ashed and some nutrients are lost by gasification. A biological activation process is needed before the soils have the ionic interchange of soluble salts. Burning can change the expansion properties of the clay particles and remove their structural water.

The short-term physical damage that occurs to the soil and residue by burning is not possi-ble to quantify and future damages are difficult to evaluate. If burning is allowed, some immedi-ate economic benefits can be obtained, but the impact of diminished production will be felt in the near future. This situation has been proven on millions of hectares in Chile and in the world. It is even more drastic in the deterioration of the fundamental soil resource, permitting the tillage to accelerate the destruction process (Crovetto, 1988).

Numerous researchers have pointed out that water conservation is greater when the soil has been unaltered and is the principal reason for production increases. Higher levels of organic mat-ter can increase water use efficiency by plants and water available for domestic or industrial use. Because conservation of residue on the soil surface is so important, I believe it is absolutely nec-essary to insist that the practice of stubble burning be eradicated in Chile. Burning not only direct-ly affects the farm and forest lands, but the smoke affects neighboring lands and cities. The large burning performed in south-central Chile during the summer produces an increase in carbon diox-ide and other gases in the lower and upper atmosphere. This not only affects our country, but the rest of the world as well.

2.6.4. Chemical Aspects

The chemical damage caused by fire is less than the physical damage. Also, more time is required to cause the damage. It has been observed that fire temporarily increases the soil's fertil-ity, benefiting in some way the next planted crop. If the heat intensity is low and there is no wind at burning time, it is possible to produce some chemical elements from the organic matter calci-nation and mineralization. Many researchers agree, however, that organic nitrogen is lost by volatilization. If the accumulation of organic material causes the fire to be too violent and the environmental conditions are poor, the rising warm air winds can remove some of the ash from the site.

Many farmers use this transitory benefit in fertility. The benefit only lasts the year of the burning because the chemical elements that are generated by the fire are eventually consumed dur-ing that period.

What really happens when the stubble is burned is that the natural process is accelerated vio-lently. This process of mineral transformation can take several years to be achieved through bio-logical organic matter decomposition. Fire achieves this drastically because the microbiology and other natural processes of the system are not involved.

The residue mineralization or calcination versus the biological decomposition can be com-pared with a car battery. The battery is built to receive a small, but constant, charge and at the same time it can deliver high quantities of energy in a very short time. When the electric system is not working well and there is an overcharging or the starter has to work more than normal, the need for energy is greater than the battery can deliver, harming its life span. Under these conditions, peak discharges can damage the battery plates, volatilizethe distilled water, and finally the capacity to retain energy is lost. Something like this happens in the residue. The quick oxidation and mineralization of the organic matter can increase the available nutrients to the plants, but only for a short time. This does not promote the colloidal activity. It can generate high losses of carbon and nitrogen because of gasification, which damages more than benefits the soil resource. Finally, we should not forget that nature does not burn residue (at least in Chile). Thanks to this natural phenomenon, we can count on the valuable soil formation process.

The lack of native forest land, the burning of crop residue, and soil tillage constitutes the death triangle that is destroying mother nature, our environment, and our own life.

2.6.5. Biological Aspects

Some specialists attribute fire as an important factor in disease and pest control. This can be effective when the disease is one of large proportions; however, we have to take into consideration that the price for this control will be very high. The natural biological agents will be affected, which can cause more ecological problems than benefits obtained from the burning.

A great quantity of insects, arthropods, and other organisms are in the upper part of the soil. They can be seriously affected by burning. Besides affecting the soil organic life, the fire deprives them of their food, which is the crop residue. This is the situation for the earthworm *(Lumbricus* sp.), an organism that is very important in soil conservation and fertility. The crop residue on the soil mitigates soil degradation and helps to achieve a better pesticide use. Because most pesticides are biodegradable under these natural environmental conditions, the existing flora and fauna are affected to a lesser degree.

Seeing a green field makes us enjoy nature. A bare or burnt soil makes us feel sad because it damages the landscape and makes it appear neglected.

2.7. Carbon/Nitrogen Ratio

2.7.1. General

At different stages of development and maturity of plants, nitrogen tissue content can vary greatly with respect to carbon. Thus, a flowering plant can have a carbon/nitrogen ratio from 15:1 to 30:1, and upon completion of its life cycle, this same plant can reach a carbon/nitrogen ratio of 80:1. This ratio is more narrow in leguminous plants and broad in gramineous plants.

When the plants reach the final stage of their cycle and produce a ripe fruit, the seeds possess a greater content of nitrogen than the plant itself. The nitrogen content fluctuates between 1.5 and 5%.

Carbohydrates are formed during photosynthesis. They combine with amino acids coming from nitrogen absorbed by the plant and form compounds called protein which is vital for animal life. Seed and fruit from vegetables extract nutrients from the main plant, and in this way receive the energy and protein required. This indicates that plants, while reaching their physiological maturity, decrease in nitrogen content, thus increasing their carbon/nitrogen ratio. This is the reason that crop residue is generally rich in fiber, cellulose, and hemicellulose, and poor in nitrogen. The carbon/nitrogen ratio of mature plants can fluctuate between 30:1 and 80:1, depending on the nitrogen level of the original crop residue.

2.7.2. Soil Carbon/Nitrogen Ratio

Nitrogen and carbon are fundamental for life, existing in virgin soil in a harmonious and stable relationship; however, the intervention of humans in the management of soil has brought about a disequilibrium, causing losses of these primary elements.

Soil carbon is in organic tissue in decomposition or mineralization pathways. It is essential for the nutrition of soil microflora and mesofauna. Here is where soil fertility develops and where carbon losses are constantly replenished through additions of fresh organic material. As already mentioned, the plow and fire are active destructors of organic matter; thus, plowed soils have low carbon levels and, consequently, low fertility.

Nitrogen is important in the decomposition of organic matter to maintain natural equilibrium. The plow and cultivation tools tend to increase the rate of the biological processes that cause losses or excessive concentrations of nitrogen that alter the chain of biological decomposition and soil fertility (Crovetto, 1986). Table 2–1 shows the carbon/nitrogen ratio of residue commonly found on an agricultural farm.

The carbon/nitrogen ratio in most agricultural soils varies from 10:1 to 12:1, which indicates that it is narrower than that of the residue, which varies from 30:1 to 80:1. Adequate nitrogen fertilization to narrow the carbon/nitrogen ratio may be important to consider in the management of residue in no-till.

The residue in contact with the soil initiates a slow biodegradation for which the microorganisms responsible require a certain quantity of nitrogen to be able to decompose it. If the carbon/nitrogen ratio is too broad, the microorganisms must extract nitrogen from the soil, producing a deficiency and competition for the established crop. This phenomenon can seriously affect the final yield of a crop. If there is not enough nitrogen, the humification process and posterior mineralization of the crop residue is delayed. For this reason, it is important to consider the additional nitrogen necessary to cover this disequilibrium.

Leguminous plants that fix atmospheric nitrogen generally have a higher nitrogen content than nonfixing plants, such as grasses and crucifers. Residue of plants that do not fix atmospheric nitrogen have a broader carbon/nitrogen ratio than leguminous plants. For example, soybean residue contains 1.2 to 1.6% nitrogen, corn 0.7 to 1.0%, and wheat straw only 0.2 to 0.6% N. This means that wheat residue requires more nitrogen for decomposition than corn. Residue that contains more than 1.5% nitrogen does not require additional nitrogen applications.

After 2 to 6 months, under favorable environmental conditions, the decomposition of organic matter in plowed soils and its transformation to humus can signify a loss of up to 50% of the carbon content as carbon dioxide that the residue contains (Allison, 1976). When nitrogen fertilizer is applied on the soil, the time necessary for the humification of residue decreases. The nitrogen applied is trapped in organic material. Only a small part of the nutrient is available to the soil and plant. Nevertheless, it will be returned to the soil once it completes humification and later mineralization.

Table 2–1. Carbon and nitrogen relation of different material (Crovetto, 1986).

Residue Type	Parts of Carbon	Parts of Nitrogen
Wheat straw	80	1
Oat or barley straw	80	1
Sorghum stalk	70	1
Corn stalk	50	1
Beans	40	1
Legume	30	1
Soil	10	1

Table 2–2. Additional nitrogen required to correct the carbon/nitrogen ratio of different residue on Chequén soils (Crovetto, 1988).

Crop	Yield	Residue	Additional nitrogen required
	tonnes per hectare (pounds per acre)		kilograms per hectare (pounds per acre)
Wheat	4–5 (3600–4500)	5–7 (4500–6250)	55–77 (49–69)
Sorghum	3–4 (2700–3600)	7–8 (6250–7200)	70–80 (63–71)
Corn	6–8 (5400–7200)	4–6 (3500–5400)	28–42 (25–38)
Beans	2–3 (1800–2700)	1–1.5 (900–1350)	5–7.5 (4.5–6.7)

2.7.3. Correction of the Carbon/Nitrogen Ratio

Correcting the carbon/nitrogen ratio in no-till is one of the greatest problems that farmers face because it requires an additional application of N beyond that of the crop requirement. The additional quantities required to correct the carbon/nitrogen ratio with different residue are indicated in Table 2–2. Nitrate nitrogen (sodium nitrate, calcium nitrate, or ammonium nitrate) should be broadcast over the residue at planting. Additional quantities of N are necessary only during the first 2 or 3 years of no-till and should be reduced to zero within the fourth and sixth years. Urea is not recommended because of the high risk of nitrogen volatilization.

2.7.4. Correction of the Carbon/Phosphorus Ratio

During the process of organic matter decomposition, microorganisms also require phosphorus for their metabolic functions. These requirements cannot be fulfilled by soil levels alone, particularly if phosphorus fixation is high. With this in mind, appropriate amounts to apply should be 50% of the required nitrogen. In soil with pH 6.5 or less, the ideal method would be to apply phosphate rock plus calcium–magnesium ammonium nitrate according to the needs of the seedlings, at least during the first 4 years. The fractions of calcium and magnesium in nitrogen fertilizer greatly aid the nutrition of microorganisms that break down cellulose and in fixation of atmospheric nitrogen by nonsymbiotic activity.

Although the chemical inputs necessary to normalize the carbon/nitrogen and carbon/phosphorus ratios require additional fertilizer costs, the long-range benefits of these inputs translate into increased and stable soil fertility levels.

The use of residue on the soil is the fundamental part of a conservation-production system in agriculture. It is the extraordinary and necessary part that complements the management of the no-till system. For all this, I can say that the human being has to give to his partner, the soil, his own food that is produced by the soil. **"The grain is for the farmer, and the residue is for the land."**

2.8. WEED CONTROL

2.8.1. General

Weed control is one of the most important aspects of crop production because weeds can seriously interfere with crop yields.

Weed control has been one of the most important agricultural concerns throughout history. Even with the help of herbicides, weed control is inadequate in many cases. Some experts point out that crop yields can be reduced from a small percentage to total crop failure as a consequence of aggressive weed competition.

Weeds are everywhere, under different climates and natural ambient conditions. Even though everyone talks about weeds, they still need to be defined.

2.8.2. Definition

Weeds are those endemic or introduced plants that can impede or damage sowing or planting. Plants that are extraneous to the established crop are considered weeds. According to this point of view, weeds occur in a pasture when livestock reject them or their nutritional contribution is poor. Under certain conditions, however, some plants that occur naturally, form an important nutritional constituent of the pasture. An example of this is *Medicago lupulina* L. This endemic legume plant is very useful for livestock feeding, but can be a serious problem in any cultivated crop.

Weeds generally develop in crop fields when preplant and postemergence soil management is inadequate and pest and disease control, irrigation management, and fertilization are poor.

Weeds can be anywhere because large quantities of seeds and vegetative structures can remain in the soil and develop when conditions are ideal. For this reason an efficient weed control program should focus on weeds that demonstrate some vegetative development as well as weed seeds that have germinated or that are near germination.

Mechanically cultivating the soil tends to reduce the presence of already established weeds; however, it creates physical soil conditions ideal for seeds on the surface to germinate. In other words, **the more the soil is mixed or tilled, the greater is the weed infestation.** This is because germination is stimulated by covering the seeds on the surface with soil and by uncovering buried dormant seeds.

Weed control should be planned, but successful achievement is not possible in 1 or 2 years. Probably more than 2 years are needed to achieve control of certain endemic weeds. If in only 1 year a few weeds reach maturity and produce seeds, then these weeds can reproduce later and can cause aggressive competition for the next crop.

Seeds of endemic, introduced, and cultivated plants alike can develop into weeds. For example, oat residue that contains seeds lost during harvest and not controlled before or shortly after planting can be a problem in a wheat crop, or wheat in a soybean crop. Thus it could be said that the concept of the weed cannot be limited to a particular plant, but also must include the type of cropping system implemented.

Weed control has never been easy, especially with a lack of adequate technology. Good control occurs only after applying a good immediate treatment followed by a complete plan.

I have mentioned that under no-till management, the types of weeds present tend to change. In a wheat rotation with lupin, a tremendous increase occurs in grasses, such as annual ryegrass, as well as other monocotyledons. The seeds of these weeds have an extraordinary capacity to germinate in residue, some without experiencing allelopathic effects (Crovetto, 1986). This is why, under no-till, effective weed control is a basic necessity.

Application methods, types of herbicides, and dosages can differ from those of conventional systems, such that in successive years of no-till farming those weeds that were targeted for control begin to disappear. This paves the way for the weeds that also were present, but, because of competition, could not manifest themselves. This is why weed control must be seasonally planned as a vital tool for successful no-till.

In no-till weed control, it is useful to cut weeds at their base at the end of spring or before their seeds mature. Livestock also may be used to control weeds. They provide an obvious bene-

Poor management of the soil can cause extensive proliferation of undesirable plants, such as thistles (Florida, 1989).

fit at no cost, but an additional mowing may be necessary after they are removed. Cutting annual weeds at their flowering stage or early fruiting stage (before the fruits reach physiological maturity) can improve their control. Timely application of a contact herbicide can assure good control of these types of weed.

Some endemic weeds, such as *Oxalis* sp. and *Rumex acetosella* L., become serious problems in annual crops when the soil acidity increases (pH decreases). Under these circumstances, applying calcium carbonate and adequate nutrients to the established crop can give good control of these biennials. *Cirsium arvense* (L.) Scop. and *Carduus pycnocephalus* L., both in the composite family, can be similarly controlled.

2.8.3. Herbicides

The farmer has a variety of chemical and cultural practices at his disposal that allow him to control weeds efficiently. The cultural practices include crop rotation, fertilization, and lime amendments. Adequate monitoring combined with experience are necessary to gain a better understanding of weeds and the specific herbicides that control them. Herbicides generally work well if the label instructions are followed and if they are applied at the appropriate stage of growth under normal conditions. The most important thing before establishing a rigid application schedule is to analyze the weed control requirements of each crop rotation with the help of a specialist.

For adequate control, the weeds must be identified by both their common and scientific names. I believe this is critical for the farmer to make the best use of herbicides. Weeds and their herbicides require continual study and experimentation on proper application rates. In this respect, it is important to support the various agronomic disciplines of agricultural research in all its various experiment stations and substations throughout the country. The technical assistance provided by commercial herbicide enterprises can help farmers. Farmers should record all the information they receive regarding weeds and their control as this information can help them improve control and lower costs.

A variety of constantly improving herbicides that meet the needs of most crops is currently available on the market. As a guide to no-till farmers, the author has classified these herbicides by their basic characteristics.

Note: The herbicides rates recommended in this book are adapted to the conditions of Chequén Farm. The reader must consult the recommendations given for their own local conditions.

2.8.3.1. Preplant Systemics

These herbicides are applied directly to the foliage. Their activity is only on the green tissues, and their active ingredient is translocated to the roots of the plants. Depending on the application rate, preplant herbicides can control most broadleaf and grass annuals and perennials. Because of these characteristics, they should only be used before planting. They are especially suitable for no-till. For example, glyphosate (Roundup, Monsanto) is a herbicide that is not active on seeds or bark and has no residual effect.

2.8.3.2. Selective Residuals

These herbicides applied to the soil before planting or preemergence to the crop act only on germinating seeds. Some important residual herbicides that are normally incorporated into cultivated soils are not very effective in no-till corn. This is mainly because the residue intercepts them, and they are rapidly photo-degraded. In Chequén, selective residual herbicides have effec-

Effect of the systemic herbicide glyphosate on established vegetative cover and weeds prior to seeding corn (1989).

Effect of grass herbicide on ryegrass in the seeding of fall lupin (1989).

tively controlled grasses in corn plantings, for example, Metolachlor (Dual 720 E.C., Ciba-Geigy) controls large crabgrass and barnyard grass. Higher rates than those recommended for average control give better results when the residue is dense.

2.8.3.3. Residual Systemics

Applied to the soil and/or over established plantings, residual systemics are very important for weed control. They act via translocation in green tissues and on pre- and postemergence seeds. In corn, for example, atrazine (Aatrex, Ciba-Geigy) controls broadleaf weeds, seeds, or seedlings.

2.8.3.4. Selective Systemics

Selective systemic herbicides act on postemergence translocation. An example is diclofop-methyl (Illoxan, Hoescht). This selective grass herbicide gives good control of annual grass weeds in wheat when it is applied in a timely manner and at adequate rates. At the highest rate shown on the label, this herbicide also can be absorbed through the roots. Picloram (Tordon 24-K, Dow), is a useful herbicide for control of broadleaf weeds in broadleaf field crops. Another good example of broadleaf weed control in broadleaf field crops is clopiralid (Lontrel 3A, Dow). This systemic herbicide acts as a weed growth regulator, altering cellular division and fundamental physiological aspects of the development of plants, such as *Anthemis cotula* L. and *Matricaria* sp. L., over which it exercises excellent control. Furthermore, it can help control common weeds in broadleaf and grass crops, such as vetch, *Oxalis* sp., chicory (*Chichorum intybus* L.), thistles, and legumes.

2.8.3.5. Systemic Grass Herbicides

These herbicides are very effective in the control of annual or perennial grass weeds in broadleaf field crops. They can be applied at any vegetative stage, but preferably while weeds are actively growing and before flowering. Examples are haloxyfop-methyl (Gallant, Dow) and quizalofop-ethyl (Assure, Dupont).

2.8.3.6. Hormonal Herbicides

The action of hormonal herbicides distorts cellular development in plants, impacting or destroying them. The herbicides have minimal residual effect and control only certain broadleaf weeds. An example is 2,4-D (Dow). Crops, such as canola, that are sensitive to hormonal herbicides should be planted at least 7 days after applying a hormonal herbicide, whether alone or in combination with other herbicides.

2.8.3.7. Contact Herbicides

Contact herbicides affect stems and leaves of annual plants during their first few months of growth. These herbicides are not translocated. An example is paraquat (Gramoxone, ICI), which is recommended for preplant use in no-till.

2.8.3.8. Selective Contact Herbicides

This type of herbicide is not common in Chile. It acts on contact, inhibiting photosynthesis and respiration of sensitive broadleaf weeds. It is absorbed by the foliage without translocation.

Excess green vegetation must sometimes be removed before applying herbicides (1989).

Effect of paraquat (a contact herbicide) on annual weeds a few days after application (1990).

Good weed control stimulates crop growth, increases yield, and facilitates weed control for future rotations (1989).

An example is bromoxyn (Bromynil, Osa). This herbicide belongs to the benzonitrile group and is used for post emergence control of broadleaf weeds in wheat, oats, and barley.

2.8.4. Preplant Weed Control

Under conventional tillage, some weed control is achieved using a plow or harrow that buries plants and seeds. Weed control is improved if two passes are made. It will be better, though, only on existing vegetative parts and not on seeds, because this method improves conditions for germination. One also should be aware that this method degrades soil structure, decreasing resistance to erosion and compaction.

Because no cultivation or soil movement occurs under no-till, weed control measures should be applied before planting by chopping or with herbicides. To facilitate subsequent weed control, it is critical to prevent weeds from forming seeds.

According to experiments at Chequén, weed control before planting should complement adequate field preparation. With the use of systemic herbicides, such as glyphosate, one can get good control if weeds are no more than 30 centimeters (12 inches) high and are uniform in growth. To achieve this condition, cut the weeds to a height of 30 centimeters (12 inches), 20 to 30 days before applying the herbicide. The cutting should be made with a chopper, an implement that also allows for removal of the excess chopped material from the field. The soil should be warm and moist enough to allow vigorous weed growth prior to herbicide application and translocation. Better and more rapid herbicide control is achieved when the pH of the herbicide mixture (water plus glyphosate) is about pH 3.5. To achieve this effect, acidify the solution. Products available on the market improve herbicide efficiency when added to the mixture.

Excessive cut and/or chopped weed material or forage should not be left on the field because it takes about 20 days of regrowth before the systemic herbicide can be applied. This is not enough time for the weeds to grow above the pasture grass or weed residue. The herbicide should only come in contact with actively growing plants, not the residue, which intercepts the herbicide intended for the weeds. When using this method, it is best to apply water soluble fertilizers on the pasture or stubble after cutting it. Nitrogen fertilization after cutting allows vigorous regrowth, which accentuates the effect of glyphosate. This fertilization shortly before planting also is important during the first years of managing the residue with no-till. Again, this procedure should be used for spring plantings in soil where weeds have over-wintered and become established as a green manure or endemic weeds.

Where weeds have been annually controlled with agrichemicals, a decrease in the weed population should be noticed after the third year of no-till. This reduction of weeds can result in residue that is insufficient to protect the soil. When this happens, a mixture of, for example, rye and clover should be planted to ensure soil protection and maintain biological activity during the semidormant winter period. This cover crop, as I've mentioned, improves subsequent weed control through allelopathic effects and competition. At Chequén, this cultural practice has given excellent results in producing corn and has improved the efficiency of capturing residual nitrate uptake and uptake of other easily leached, soluble nutrients.

Paraquat, when followed by post emergence weed control, has given satisfactory results in fall plantings in annual weed residue. Generally, it works well on newly developing annual weeds. Paraquat acts on perennial plants by burning their leaves and stems, setting back their growth. Its use as a preplant herbicide in the spring is not recommended if the weeds were established over the winter. Its effect is improved when applied at night and when mixed with diuron. Because of its greater toxicity, herbicide applicators should be extremely careful when using paraquat, taking serious measures to avoid inhalation during the application and/or contact with the skin.

In no-till, weeds can be controlled before planting, though this is not enough. Biannuals and perennials are difficult to control with postemergence herbicides, especially when initiating no-till.

2.8.5. Preemergence Weed Control

Weed control immediately after planting and before crop emergence can be a useful tool in certain crops. In winter wheat plantings, diuron (Ustinex 80% WP, Bayer; Karmex 80 PM, Shell) produces good results on difficult-to-control weeds in the seedling stage. The herbicide label should be strictly followed, considering soil moisture, level of organic matter, and phenological stage of the seedlings. Furthermore, the wheat seeds should be planted as deep as possible, at least 5 centimeters (2 inches).

Some preemergence herbicides can control broadleaf weeds in winter lupin. The herbicides should be used correctly to avoid phytotoxic damage to the lupin. Linuron (Afalon, Hoechst) has demonstrated acceptable broadleaf weed control in this crop. In Australia, good preemergence weed control in *Lupinus angustifolius* L. has been obtained with 2 liters per hectare of simazine, a herbicide in the triazine family. This herbicide adequately controls broadleaf weeds in this crop (Gilbey, 1982). It was used for the first time at Chequén in 1990 just after sowing lupin. We had good results on a variety of weeds, including grasses.

Preemergence weed control with herbicides applied on residue is not always efficient because they can damage any germinating seed. Therefore, all planting should be done as deep as possible within the allowable depth for the particular seed in question. These herbicides should be applied in doses based on research, depending on the soil, moisture, and content of organic matter. They should be applied immediately after planting before some emerging seedlings can be damaged.

Generally, preemergence herbicides are applied when postemergence herbicides specific to the crop are not available. The problem is with legumes, cruciferas, and other nongraminaceous plants, for which preemergence herbicides are less efficient. At Chequén, we observed a strong residual effect of linuron when applied in corn after emergence for control of *Digitaria sanguinalis* (L.) Scop. Six months after applying this herbicide at a rate of 2.5 liters per hectare, linuron reduced oat germination to the point where we had to replant 30 days later. The constant progress in herbicide development will soon provide farmers with more selective, postemergence, systemic residual herbicides for broadleaf crops.

Residual herbicides can have negative effects on established crops. If soils are sandy or coarse-textured, it may be possible to reduce the application rate recommended by the manufacturer. Conversely, clay soils and soils that have high organic matter content and low pH may require the highest application rates allowed by the label. This is because the colloid minerals and organic material actively inhibit the residual effect of these herbicides.

2.8.6. Postemergence Weed Control

Weed seeds continue to germinate after crop emergence. They need to be controlled at that time with appropriate herbicides. Weed control in small grains has become more effective every year with the introduction of new products into agricultural markets. Control of grasses in wheat is always effective when the product labels are strictly followed. Selective grass herbicides don't affect most varieties of wheat, but they can damage or destroy oats. A wide variety of herbicides, from systemic selective to systemic residuals and hormonals, are available to control broadleaf weeds in small grains.

A series of herbicides for no-till wheat and oats has been tested at Chequén with poor effects on many common weeds, such as *Anthemis cotula* L. and *Raphanus sativus* L. In 1988, tests were made with methyl-amino- carbonyl-sulfonil-benzoate (Ally, E.I. du Pont de Nemours & Co., Inc.) at rates of 8 grams per hectare. This herbicide provided good control of *Anthemis cotula* and *Oxalis* sp. L. and acceptable control of *Raphanus sativus* and *Polygonum aviculare* L. This selective systemic herbicide with residual effects, belonging to the group of sulfonyl ureas, apparently is not absorbed by surface soil organic matter as are most of the commonly used herbicides. Care should be taken with the residual effect of this herbicide as it can damage broadleaf plants even a year after application (Table 2–3).

As previously mentioned, a wide variety of herbicides can give good results if they are used correctly. I feel it is very useful for farmers to experiment in small areas to evaluate their results. Innumerable factors act on the efficiency of herbicides. They include soil type, pH, organic matter, vegetative cover, precipitation, and the growth stage of the weeds to be controlled. Other, more direct factors, include application on sloping land, nozzle type, cleanliness of main and nozzle filters, accuracy of application, formulation of herbicides and their dispersion in water, adjuvants, type of water, pH of the herbicide mixture, wind velocity, ambient temperature and humidity, overlap of the spray boom between passes, and speed of the equipment. These factors, which seem like small details, are important in executing weed control.

One could write volumes on this topic; however, this is not recommended, because I believe it is important not to give recipes that work for some farmers and not for others. I don't think it is wise to rely only on results of some of the practical observations made on Chequén.

Up until 1988, broadleaf weeds were a serious problem in corn even with applications of 2.5 liters per hectare of atrazine. This herbicide wasn't effective in controlling germinating weed seeds. This showed a clear deficiency in the product and resulted in consequent damage to the crop.

In 1989, calcium carbonate was applied to the soils at Chequén to improve the soil pH, which had been decreasing from urea applications until it had reached pH 5.77. Applying 1500 kilograms per hectare (1340 pounds per acre) of this amendment significantly raised the pH to 6.91. Surprisingly, that year the broadleaf weeds common in corn disappeared, leaving the soil practically clean. It became evident, then, that the act of raising the soil pH improved the residual effect of atrazine applied preemergence to the soil surface. It should be noted that some herbicides, including those in the triazine group, are inactive in low pH soils (Tisdale et al., 1985).

Table 2–3. Herbicides used with corn.

Herbicides	Commercial product	Water
	kilogram/hectare	liters/hectare (gallons/acre)
Preplant†		
Roundup	2.5	350 (37)
2,4-D (80% ester)	1–1.5	
Preemergence§		
Atrazine	2.5	350 (37)
Paraquat	2	

† 15–20 days before planting.
‡ Use only for weeds resistant to Roundup: For example, *Trifolium, Medicago,* and *Erodium.*
§ For adequate control of *Digitaria sanguinalis* and *Echinochloa crusgallis* (L.) Beauv., add 2 liters per hectare of preemergence metolachlor.

Table 2–4. Herbicides used with winter wheat planted into stubble of oats, triticale, lupin, and canola.

Preplant herbicides†	Commercial product	Water‡
	liters/hectare	liters/hectare(gallons/acre)
Preplant		
Paraquat	2	200–300 (21.4–32.1)
Postemergence§		
Ally tank mix	8 g	200–250 (21.4–26.8)
MCPA 750 amine	800 g	

† Weeds should be between 30 and 60 days old.
‡ If weed growth and density are low, use less water with preplant herbicides.
§ It may be necessary to apply a selective grass herbicide at least 7 days before or after applying Ally.

Good control of annual ryegrass and other problem annual grasses was observed throughout the early spring weed control period in soils scheduled for corn planting. We even ended up hand weeding, but the corn was weakened from competition by the weeds.

Then in 1989, exceptional control of these weeds was obtained using the same herbicides and application rates as before. The difference can be the greater amount of water applied, 350 liters per hectare (37 gallons per acre) of water versus 250 liter per hectare (27 gallons per acre) as was the usual application rate recommended by the manufacturer. I give importance to this fact in as much as systemic herbicides should cover the plants' leaves and stems as much as possible. Only in this way can the green tissues efficiently translocate the active ingredient and destroy the weeds. On the other hand, even though contact herbicides aren't translocated to the plant roots, heavier water applications can still improve herbicide action. In any case, one should consider that over application of water is only recommended when weeds completely cover the soil and are actively and vigorously growing (Table 2–4).

On Chequén corn fields when weeds are still brownish green 2 weeks after applying glyphosate, we have applied paraquat and atrazine as the last weed control. This has given better preemergence control. Paraquat controls some plants still in a tentative vegetative stage, and atrazine helps control broadleaf weeds in addition to having the residual effect on controlling seeds that are about to germinate.

2.8.7. Weed Control Implements

It has already been noted in Section 1.6.2 that sicklebar mower and disk cutters, mulchers or choppers, or horizontal cutters are adequate implements for controlling annual weeds, simply because they can prevent seed formation. On the other hand, most weeds will regrow when the crops need to be clean. For this reason, weed control implements are a good help especially when they are used during the planting or drilling period. Residue left after these cuttings provides efficient soil protection and helps to improve no-till management.

Brazilian farmers who continuously practice no-till successfully use a tractor-pulled roller flattener, that is three rollers in triangular formation. These rollers have sharp protruding blades arranged in parallel lines 15 centimeters (6 inches) apart, such that they flatten and crush the weeds in the same pass. This mechanical action is useful in controlling weeds or green cover crops, especially cover established in the autumn or winter. Oats planted simply to protect and improve the soil during this period are easily controlled with this implement, which crushes the oats, almost cutting the stem several times along its length. To get good weed control, the oats should have reached at least the panicle stage before the roller flattener is used. Once they have

The Delavy Roller was used to flattenen and control weeds and oats that were established as a green covercrop prior to no-till planting (1990).

A roller flattener was used to flatten headed oats (Passo Fundo, Brazil, 1990).

started to head, oats will not regrow or reseed. This implement should only be used on plants that are 40 to 100 centimeters (16 to 40 inches) high. The tractor should pull the implement as rapidly as possible.

The high density of dead vegetative cover left on the soil by the roller flattener can generate positive allelopathic effects that help control weed seed germination. The roller should be used 20 to 30 days before planting. If regrowth occurs, adequate preplant herbicides should be applied. The no-till planter should be pulled in the same direction as the roller flattener to facilitate penetration of the slicing and planting coulters.

2.9. HERBICIDES IN SOIL MICROBIOLOGY

The constant increase in herbicide use is an international concern. The no-till system has many detractors simply because weed control is done by chemical means, which supposedly can contaminate or damage the ecology of the soil and human health.

It is evident that weed control in no-till must be performed with herbicides because the system doesn't allow soil disturbance by cultivators or similar implements. No-till became possible once herbicides were able to control weeds, and without them, we would still be waiting for the no-till arrival.

The interaction between humus and herbicides should be noted in soils with normal or high levels of organic matter. Organic compounds can influence the persistence of herbicides in the soil, inactivating them or protecting them from further microbial degradation. In fertile soil, the occurrence of enzymatic activity is high and is associated with the level of organic carbon in the soil and the presence of an active and diverse microbial population (Zunino, 1982).

When we first began practicing no-till at Chequén, we didn't study the behavior of herbicides, especially whether they could affect soil respiration and consequently the soil microflora. Later, in 1981, Francisco Gavilán, a biologist at the University of Concepción, studied the three most commonly used herbicides at Chequén: glyphosate, atrazine, and paraquat. This study examined the behavior of algae, azotobacter, fungi, bacteria, and actinomycetes in soil samples obtained from no-till fields that had received applications of each of these herbicides for 3 years. The results obtained in this study are presented in the following section.

2.9.1. Glyphosate

It was established that glyphosate, also known as Roundup, which is considered important for control of weeds prior to planting, did not affect the population of soil fungi. It even increased the soil bacterial and actimomycetes populations compared with those of the control group. This herbicide slightly inhibited development of green algae during the first 5 days after application. Glyphosate promotes microbial activity in the soil by serving as a nutritious substrate of carbohydrates in its final stage of biodegradation (Gavilán, 1981). The results obtained from this study are shown in Table 2–5.

The most important effect of this herbicide, even at rates as high as 6 liters per hectare is far from affecting the development of the soil microflora. It significantly promotes certain microorganisms. Glyphosate continues to be an exceptional herbicide for general control of weeds before no-till planting, and it is really one of the most important tools for managing this system.

Table 2–5. Effect of Roundup applied at 6 liters per hectare on certain microbial populations colonies in a sample of no-till soils from Chequén (Gavilán, 1981).

Microbial group	Treatment factor	Colonies formed per gram of dry soil			
		Time of incubation (days)			
		0	5	10	20
Algae ($\times 10^2$)	Control	4.00	6.75	6.00	5.75
	Roundup		2.75	5.75	4.75
Azotobacter sp. $\times 10^4$)	Control	3.47	56.75	8.52	9.35
	Roundup		74.25	9.35	9.00
Fungi ($\times 10^3$)	Control	3.50	67.00	85.00	107.00
	Roundup		63.25	85.25	113.00
Bacteria ($\times 10^5$)	Control	55.70	65.25	18.40	9.07
	Roundup		90.00	215.00	90.25
Actinomycetes ($\times 10^5$)	Control	54.50	7.47	6.12	8.00
	Roundup		6.82	54.00	50.70

2.9.2. Atrazine

Atrazine has been used at Chequén since 1975. This systemic residual herbicide, used pre-emergence and postemergence, helped control weed seeds and noxious plants in corn. In 15 years of use, no adverse effects have been noted except the effects on algae and actinomycetes during the first 20 days of incubation; however, it did not affect the azotobacter, fungi or bacterial populations. Table 2–6 shows the effect of atrazine on part of the soil microbiology.

2.9.3. Paraquat

During the entire incubation period, paraquat strongly affected the algae population, decreasing its activity by more than 50%; however, the rest of the soil microflora studied didn't show any significant changes. Table 2–7 shows the effect of atrazine on part of the soil microbiology.

Table 2–6. Effect of atrazine at 3 liters per hectare on certain microbial population colonies in a sample of no-tilled soils from Chequén (Gavilán, 1981).

Microbial group	Treatment factor	Colonies formed per gram of dry soil			
		Time of incubation (days)			
		0	5	10	20
Algae ($\times 10^2$)	Control	4.00	2.00	21.50	6.00
	Atrazine		0.50	5.50	2.70
Azotobacter sp. ($\times 10^4$)	Control	3.47	5.40	3.30	0.85
	Atrazine		6.00	2.80	0.92
Fungi ($\times 10^3$)	Control	3.50	6.90	16.00	9.00
	Atrazine		7.10	19.00	7.40
Bacteria ($\times 10^5$)	Control	55.70	14.00	14.20	12.77
	Atrazine		14.10	15.70	15.92
Actinomycetes ($\times 10^5$)	Control	54.50	5.60	11.87	8.77
	Atrazine		4.10	6.45	5.65

Table 2–7. Effect of paraquat at 3 liters per hectare on certain microbial population colonies in a sample of no-tilled soils from Chequén (Gavilán, 1981).

| | | Colonies formed per gram of dry soil | | | |
| | | Time of incubation (days) | | | |
Microbial group	Treatment factor	0	5	10	20
Algae	Control	4.10	10.50	9.50	19.00
($\times 10^2$)	Paraquat		3.75	4.00	8.25
Azotobacter sp.	Control	3.47	3.80	4.70	1.35
($\times 10^4$)	Paraquat		5.45	5.35	1.70
Fungi	Control	0.35	1.20	1.70	3.30
($\times 10^3$)	Paraquat		1.08	1.97	1.72
Bacteria	Control	5.57	7.80	7.40	7.90
($\times 10^5$)	Paraquat		8.56	7.02	5.80
Actinomycetes	Control	5.45	16.30	10.50	0.28
($\times 10^5$)	Paraquat		15.10	12.80	0.38

According to the author of this study: The herbicides that cause the least ecological effect to the soil microflora, in order, are glyphosate, atrazine, and paraquat. One could add that systemic and contact herbicides are retained on the surface of leaves and stems, and very little of these herbicides reaches the soil.

In no-till, residual postemergence herbicides fall on green vegetation, residue, or the small amount of soil that the planter exposes. As a result the chemicals are exposed to the direct action of solar radiation and/or volatilization causing a decrease in herbicidal effect. One method of minimizing this situation would be to irrigate soon after application to move the herbicide into the soil. If this is not feasible, one should increase the manufacturer's recommended water dosage by 50% compared with soils under conventional tillage.

Herbicides are a fundamental help in managing no-till. Nevertheless, their continued use demands analysis of their effects on soil microbiology and other natural parameters. At Chequén, results with the herbicides mentioned indicate that the residuals can have the greatest effect on the biological equilibrium. These herbicides remain in contact with the organic material near the surface of the soil.

Herbicides in no-till cause much less damage to the soil microbiology, mesofauna, and humans, compared with the effects with conventional tillage.

As already expressed, ideal control of weeds in no-till is in the preplant stage where products or mixtures are applied to existing vegetation. To complement an efficient control, as stated previously, the weeds should be prevented from forming fruits and seeds so that use of residual herbicides can be reduced.

It is important to point out that the largest human population in the world's history must be fed with sufficient, high quality foods. Herbicides are necessary today to achieve this. **The selection must be between herbicides or the plow, erosion, and a shrinking food supply. You make the choice.**

2.10. Fertilization in No-till

Selke (1968) points out that "the basis of all agriculture is the soil that is in the weathered surface of the Earth's crust." As a farmer and through experience obtained at Chequén, I fully share in this assertion and have proven it to be true. Soil, in addition to furnishing the substrate and physical sustenance to the plants, also provides storage of water and nutritive substances. It is of major importance from the perspective of plant use and fertilization.

In plant production, soil fertility is one of the most important factors. The addition of materials to the soil–plant–water system positively influences the development, yield, and quality of the harvest products. Proper nutrient management can change the natural productive parameters of the soil. This fact is important when the natural soil fertility is to be improved. The fertility of a compound is not limited to only nutritional qualities, since in certain circumstances indirect benefits can result. For example, nutrients can be transformed into compounds which the plants do not have direct access, or they can improve the physical, chemical, and biological characteristics of the substrate (Selke, 1968). According to my experience, I share this concept in the sense that, if one wishes to use fertilizer rationally, it is necessary to know the fundamentals of plant nutrition and soil physiology. This can allow the application of concepts of productivity to specific circumstances in each case.

The use of fertilizers has been increasing since World War II. The major reason is that soils have gradually lost their natural fertility as a result of poor management practices that have resulted in erosion, salinization, leaching, volatilization, and extraction of nutrients by plants. Also, fertilizer has increased to meet the need for increased yields.

Although there are many detractors and critics of chemical fertilizers, I regard fertilizer as necessary in a world that demands more food each day while the soil is becoming less and less productive. I firmly believe that together with the notable advances in plant genetics and new efficient herbicides, fertilizers form a fundamental third part of a triangle. This triangle is responsible for the feeding of humanity.

Many fertilizers are available on the market and a number of varieties exist. The choice must be made on cost and agronomic efficiency, considering the soils physiological parameters. Farmers must know the fertility levels of their soils based on analyses made by specialized laboratories with consideration of the nutrient needs of the crop to reach the expected yield. Observing these two conditions can produce good yields without decreasing the potential productivity of the soil.

The great majority of cultivated crops in Chile require fertilizers, principally nitrogen, phosphorus, and potassium, which are basic chemical elements in plant nutrition; however, in no-till, crop needs can change, not because of a lower or higher fertilizer requirement, but because of varying availability of soil nutrients. This clearly explains why the traditional parameters in nutrition change when the soil is not tilled and the residue is integrated into the productive management system. Therefore, when farmers plan their nutrient budget to determine how much fertilizer is required by a certain crop, generally they calculate the crop needs based on desired production levels without considering the basic nutritional needs of the soil to maintain productivity.

One of the most important observations in crop fertility at Chequén has been precisely a change in the traditional norms of plant nutrition. This has caused me to conclude that in **no-till not only must we worry about crop nutrition, but also soil nutrition**. In accordance with this point of view, the farmer must know the fertility levels of the soil before planting, and in the following years, observe changes comparatively with the previous years' analyses. This is the time to correct any deficiency or excess.

With the passing of the years and carefully observing the advances in plant production and the different soil analyses of Chequén, I can say that two fundamental parameters exist in the use of fertilizers and final yield of the crops. I am referring to the organic matter levels and the soil reaction (pH).

2.10.1. Organic Matter and Nitrogen

The natural mineral components of the surface soil provide little nitrogen. Soil nitrogen basically comes from the atmosphere and also, as a product of biological recycling from organic matter in the process of degradation and transformation into more complex humic matter.

Plants cannot assimilate nitrogen directly from the atmosphere with the exception of leguminous and some other genera of plants. In some cases this chemical element, in the form of urea, can be absorbed by foliar parts of the plant. Normally, the soil nitrogen that benefits plants comes from microbial activity. The symbiotic or free-living microbes act on nitrogen-containing organic compounds that are rapidly transformed from organic nitrogen to the nitrate form by ammonification followed by nitrification. Nitrogen in this form contributes significantly to plant nutrition and good yields (Demolon, 1967).

Besides the potential obtained by biological means, it also is possible to supply nitrogen to the soil in regions where electrical storms are abundant and atmospheric nitrogen can be precipitated with rainfall.

Except for some soils that have slow mineralization of organic matter, such as the trumaos soils, there is a strong relationship between organic matter and nitrogen in the soil.

The nitrogen reserves in the soil can rarely provide adequate quantities of this element to the crop necessary to obtain high yields. For this reason, nitrogen constitutes a limiting factor if the requirement of the seedlings is not supplemented during the period of active growth. Nevertheless, under certain conditions, generally when the soil crop system has been well managed, the soil itself can supply up to two-thirds of the nitrogen needs. The rest must be supplied as commercial or organic fertilizers (Demolon, 1967).

Agronomic research points out two essential aspects that should be considered in relation with the liberation of nitrogen from organic matter. One is the high percentage of this chemical element in the organic substrates of a soil that is not plowed. In one year of tillage soil nitrogen decreases to the level of a soil that has been plowed for several years. The other aspect is that the organic matter content decreases progressively in a soil that is tilled annually.

In no-till, the contribution of carbon and nitrogen released by the decomposition of organic compounds and the recently-added crop residue is slow because of the ever increasing reserves of organic matter. The residue of recent crops, however, are decomposed more rapidly than the soil organic matter itself.

Broadbent, cited by Thompson (1965), mentioned that the addition of residue speeds up the decomposition of the organic matter of the soil; however, this does not mean that the total percentage of nitrogen and carbon will decrease. To the contrary in no-till, the fresh residue will increase the biological activity and, in the case of trumaos soils, increase the population of actinomycetes, but reduce the population of other microorganisms. This is caused by the active presence of allophane (aluminosilicate) in the trumaos soils.

In no-till, when the residue is properly managed, the ecological demands of the soil are approximately met. This soil management differs greatly from the traditional system and has the virtue of increasing organic matter levels and achieving a better use of the applied and soil-available nutrients. Almost all soils (except the trumaos from the Andean foothills) that contain organic matter levels greater than 4% are reasonably fertile. Soil organic matter content of less than 1% is an indicator of a poor soil. I am trying to explain that a **direct proportional relationship exists between soil fertility and the level of organic matter**. The exception is the soils derived from recent volcanic ash with high content of allophane.

No-till soil has a formidable advantage over soil that is exposed to tillage through its capacity to retain and transform nutrients. Organic matter is the base that supports soil microbiology. A gram of organic soil can contain millions of beneficial bacteria and fungi as well as many other organisms that transform the crude organic matter into humus and subsequently into useful nutrients for plant nutrition.

Crop residue is the base for enriching the soil besides protecting it from erosion. A soil with abundant well-managed residue on the surface increases the mesofauna and microorganism populations because of the suitability of the environment and energy available for these organisms to proliferate. On the other hand, it is useful to point out that applied fertilizers in an environment such as that already mentioned are used by plants in a more efficient way. Where additional fertilizer is applied just before the moment of rapid plant development, the benefit increases. A soil with an adequate organic matter level can help retain nutrients, minimizing the losses by leaching and volatilization. **Organic matter together with mineral colloids are very good nutrient accumulators.**

Agricultural production has been historically based on the principal of nutrient accumulation, a statement that obviously has a limit. Actually, soils in general suffer a severe depletion of nutrients because of the permanent decline of organic matter levels, a product of an agriculture based on extraction. This classic phenomenon is prevalent worldwide. The soil organic matter levels in developing countries have been depleted to such an extreme that they can only support, in the best of cases, a subsistence agriculture. **In developed countries the low levels of soil organic matter is not reflected in soil fertility levels because the severe deficiency in nutrients is solved by a dependency on agrichemicals, especially commercial fertilizers.**

In no-till, management of residue is truly interesting and crucial. When properly planned and implemented, it results in positive benefits on soil microbiology and fertility.

2.10.2. Fixation of Atmospheric Nitrogen in the Soil

2.10.2.1. Symbiotic or Associative Fixation

A good level of soil fertility can be achieved in the short-term by efficiently managing the crop residue. Under these conditions, the soil microbiology can be reinforced and, with this, the natural mechanisms of nitrogen fixation that most leguminous plants possess. This fact is significant economically because only a minimum use of commercial nitrogen fertilization is required.

Bacteria of the genus *Rhizobium* sp. can enter the roots of leguminous plants and form nodules, some more than 12 millimeters (0.5 inches) thick. These nodules are processors of atmospheric nitrogen, which they fix for the plants metabolic system. Experience has demonstrated that a specific grade of bacteria is adapted to each legume species that maximizes the infectant power on the root and capturing of nitrogen. This depends on the natural soil conditions. To obtain the maximum advantages of this symbiosis, trials have been conducted to introduce bacteria strains by adding pure cultures of specific bacteria directly in soils that do not currently contain them. The inoculation of legumes constitutes a true microbial fertilization, which is achieved by impregnating the seed surface with a bacterial suspension in water. The resulting benefit is increased yields and lower costs. These microorganisms, called symbionts, live in strict dependence and relation with the inoculated plants by natural or artificial means.

Nitrogen contributions from the Rhizobium activity can be so high that leguminous plants do not require additional nitrogen fertilization to produce good yields of protein-rich grains.

Consideration should be given to two aspects of the symbiosis of nitrogen. Demolon mentions, "One should consider symbiotic nitrogen fixation as a natural process that breaks the progressive nitrogen depletion of soils in such a way that this process must be a maximum activity." He also adds that: "It has been observed that in the absence of nodulation, nitrogen fertilization

Nitrogen-fixing nodules on roots of *Vicia dacycarpa* that have been stimulated by no till (1989).

Nodulation on roots of sweet lupin (1989).

ensures results as good as symbiotic fixation" (Demolon, 1967). This means that in a soil rich in nitrogen, symbiosis is suppressed and transforms, for example, the grass–legume association into competition. The advantages of a grass–legume association are not manifested in these soils.

The fixation of nitrogen shows the importance of using rotations that include a legume, as much for the production of nitrogen as for the stimulation of the soil microbiology. No-till stimulates beneficial action of Rhizobium by not disturbing the soil structure and by leaving residue on the soil. These two actions establish a basic nutritional substrate for the survival and natural reproduction of microbes.

2.10.2.2. Asymbiotic Fixation

No-till and organic matter promote free-living or asymbiotic microorganisms that do not require a host plant to exercise their nitrifying action. These microorganisms include *Azotobacter, Spirillum,* and blue-green algae.

The culture of *Azotobacter* sp. in the Biotechnology and Microbial Ecology Laboratory of Jorge Molina in Buenos Aires has delivered spectacular results with respect to nitrogen fixation. The quantity of fixed nitrogen can fluctuate between 1535 and 2715 kilograms per hectare (1370 to 2425 pounds per acre). This occurred when an adequate quantity of phosphorus, potassium, calcium, and sulfur existed in the soil that was covered with corn residue and contained earthworm casts (Lavados, 1989).

These asymbiotic nitrogen processes are generally spontaneous and natural, so it is not relevant to inoculate the soil with nitrogen-fixing bacteria. It is more important to create natural conditions in the soil so these phenomena occur (Primavesi, 1984).

Soil porosity is fundamental for achieving a good biological activity. A no-till system maintains the soil pore spaces produced by roots that have fulfilled their vegetative cycle. This improves the porosity and with it the oxygen levels in the soil rhizosphere. Moisture retention increases, which generates an adequate medium for the proliferation of microorganisms.

The soil, besides constituting a storehouse of nutritional elements for plants, also attempts to satisfy the nutritional requirements of microorganisms present. The microorganisms are not only competing for nutrition, they also are behaving as associated groups of organisms that mutually benefit themselves, the soil, and plants. Plants, through photosynthesis, provide energy and metabolic carbon to the microorganisms, which in turn decompose plant residue, liberating nutrient elements for the plants. An active and flourishing microbial population is a good indication of fertile soil and is necessary to maintain the fertility level.

2.10.3. Organic Matter and Phosphorus

Phosphorus is a vital chemical element in plant nutrition, playing an important role in enzymatic reactions. It is a constituent in the cell nucleus and, therefore, essential for cell division and for the development of meristematic tissues (Russell and Russell, 1967).

The phosphorus in the soil is absorbed by the plants primarily in the monovalent form, H_2PO_4, known generally as phosphate. Large quantities of phosphate are in organic matter, although it is not a function of the organic matter accumulation in the soil (Thompson, 1965). Espinoza states that in all Chilean volcanic soils analyzed by different specialists during 20 years of research, the content of organic phosphorus was greater than that of inorganic phosphorus. This pointed out that more investigations were needed on this subject (Espinoza, 1975).

Thompson points out that phosphates form insoluble inorganic components that are adsorbed tightly by the colloidal particles of the soil in a very slowly exchangeable form. This

Comparative trial in Petri plates to determine the presence and activity of *Azotobacter* sp. Left plate: Conventional tillage soil. Right plate: Treated with phosphates.

Comparative trial in Petri plates to determine the presence and activity of *Azotobacter* sp. Left plate: No tillage soil. Right plate: Treated with phosphates (Courtesy Alejandro Cariola, Argentina, 1990).

slowly exchangeable form is not very important for direct plant nutrition since the roots take-up soluble phosphorus from the soil solution at a much higher proportion than they obtain by ionic interchange (Thompson, 1965).

Emesley shows that the quantity of phosphorus available to the plant and that present in soil solution, in general, is too low, averaging less than 1 part per million, but could reach approximately 5 parts per million. At the same time, however, the total phosphorus content of the soil can range from 1500 to 2000 parts per million, expressed as P_2O_5. A narrow relationship exists between the total phosphorus and the organic matter providing evidence that a very significant quantity of phosphorus is not available to the plant. This unavailable fraction is called fixed phosphorus, or reserve phosphorus (Emesley, 1982), and can be in the soil in the following forms:

(a) Mineral phosphate, present in the original rocks of the soil,
(b) Metallic phosphates, as metallic salt precipitates, principally in the form of calcium, iron, and aluminum phosphates,
(c) Organic phosphates, also precipitated as metallic salts, and
(d) Absorbed phosphates in interaction with metallic ions of the surface of soil active particles, which correspond to phosphate interchange and could contribute to mineral fertilization

Phosphate fertilizers normally used by farmers can form part of any of the above mentioned forms. Available phosphorus in the soil is found in equilibrium with the fixed phosphate, as shown in Fig. 11.

In this figure, upon incorporating soluble phosphate in the soil as fertilizer, the equilibrium tends to move toward fixation. On the other hand, as the organic phosphorus decomposes, it is converted to the soluble, plant-available form. The phosphorus concentration in the soil solution increases, thereby moving the equilibrium toward the middle until the original level is reestablished with time. In the same way, as the plant absorbs phosphorus from the soluble form, the phosphorus concentration is decreased in the soil solution. The weakly adsorbed phosphorus is then displaced to maintain the concentration of soluble phosphorus. If the microorganisms tend to absorb soluble phosphorus, then the equilibrium will tend to maintain the concentration in the solution (Thompson, 1965).

The factors that affect immobilization and mineralization of phosphorus in the soil are associated with the present microbial activity. Although mineralization and immobilization are reversible processes, they are not the reactions in the true sense of chemical equilibrium.

According to conducted studies, around 75% of the phosphate applied to the soil as chemical fertilizer ends up as adsorbed phosphate by the soil particles 1 year after application. This phenomenon is initially fast, but later slows (Emesley, 1962; Espinoza, 1975). This agrees with what happens in the trumao soils that have a high content of amorphic clays. The monocalcic phosphate fertilization fixation is excessively high compared with soils with crystalline clays. Chemical fertilization is not the only option to help crops obtain the required phosphorus. Some biological organisms stimulated by the presence of crop residue in the soil generate a biotic environment that can mobilize phosphorus and improve plant nutrition.

Among the microorganisms useful in the movement of phosphorus are *Mycorrhizae*, *Aerobacter, Pseudomonas*, and especially *Megatherium*. These microorganisms can produce an enzyme called phosphatase, which is the chemical–biological basis for plants to assimilate phosphate, starting from organic or slightly soluble phosphate.

Dr. Fernando Borie, Professor of the University of the Frontera at Temuco, detected in soil analyses at Chequén that the phosphorus levels were very high with no-till, reaching almost double those of the trumao soils without no-till. The no-till soils are high in phosphatase. Dr. Borie

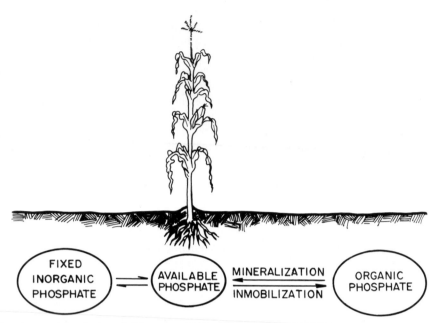

Fig. 11. Phosphorus relation in the soil (Thompson, 1965).

adds that the number of fungal spores is relatively high in no-till compared with other soils of the world.

He also detected some fungi that can form a symbiotic relationship with mycelium that invade the plant roots in a complex called mycorrhizae. The plant provides carbohydrates to the fungus, and the fungus provides inorganic nutrients extracted from the soil. It has been proven that a great majority of plants are closely dependent on these endotrophic fungi. Mycorrhizae have no special ability to extract phosphates or other nutrients directly from the soil; however, their bene-ficial action is based on the production of phosphatase enzyme that acts on organic compounds, allowing phosphorus to be liberated and available for plants. In this sense, the literature shows that various types of mycorrhizae exist. Some are more effective than others in the stimulation of plant growth (Emesley, 1982).

In other microorganisms, called vesicular-arbuscular mycorrhizae (VAM), the quantity of spores in soils under no-till have increased considerably over the years. At the same time, a cer-tain relationship can be established between the increase of VAM spore population and the increase in assimilable phosphorus by the plants, as will be discussed in Chapter 4. According to Dr. Fernando Borie, a complementary relationship exists between the enzyme phosphatase and VAM. Both are responsible for a greater availability of phosphorus to plants.

Therefore, no-till generates richness in the soil. The microbiology is stimulated by the use of crop residue, which subsequently helps the productive capacity of the soil.

2.10.4. Soil pH

Soil reaction, from the chemical point of view, can be acid, neutral, or alkaline, a charac-teristic that can be measured by pH. The reaction is acid when the hydrogen ion (H^+) is predom-inant in the soil solution. It is alkaline when the hydroxyl ion (OH^-) is predominant and neutral when hydrogen and hydroxyl ions exist in equal concentrations.

The pH scale is used to express the relative concentration of hydrogen ions in soil solution. The pH is a measure of acidity. The pH scale ranges from 0 to 14. In natural conditions for living plants, the pH can range from 4.0 to 10.0. Acid soils generally have a pH below 7.0 and alkaline soils a pH above 7.0. The pH scale that Thompson (1965) shows is an important reference in the study and care of soils. In its simplest expression, soil acidity is the result of the substitution of bases by hydrogen in the soil exchange complex. The alkalinity is the result of the accumulation of bases or cations.

Organic matter is one of the most relevant factors in the agronomic management of soils. The pH also is important in plant nutrition because an acid or alkaline soil can affect the microbiology related with soil fertility and also the nutrient available for plants. The pH effect on plant development can be expressed in relation to nutrition since the quantity of available nutrient elements is strongly influenced by the pH.

pH Scale of Soils (Thompson, 1965).

5.5	6.0	6.5	7.0	7.5	8.0	8.5
Very acid	Medium acid	Weak acid	Very weak acid	Very weak alkaline	Weak alkaline	Medium weak alkaline	Very alkaline

Because the pH is a logarithmic value of the reciprocal of the hydorgen ion concentration, the soil pH must be known because the physiological phenomenon performed within each specific pH zone can be influenced by small changes in the pH. This means that if a soil has a pH of 6.0, its hydrogen ion concentration is 10 times higher than that of a soil with pH of 7.0. A soil that has a pH of 5.0 would have a hydrogen ion concentration 100 times higher than that of a neutral soil. Subsequently, if the soil had a pH of 4.0, the concentration of hydrogen ions would be 1000 times higher (Thorup, 1984; Thompson, 1965).

It is clear that an appropriate soil pH can have greater influence in crop development than fertilization itself.

A soil with low pH (below 5.5) can cause aluminum (Al^{+3}) and manganese (Mn^{+2}) to become soluble, reaching solution concentrations that are toxic for crop development. Concentrations as low as one parts per million of soluble Al^{+3} and 100 parts per million of soluble Mn^{+2} can inhibit the vegetative development of plants (Garavito, 1979).

This fact is particularly important in trumao or volcanic soils because the soil acidity not only interferes in the phosphorus availability to plants, but also stops the normal development of the biological agents responsible for the processes of mineralization of the soil organic matter.

Phosphorus solubility notably decreases below a pH of 6.5 and above 7.5. The optimum range of phosphorus availability is between a pH of 6.5 and 7.5 (Thompson, 1965).

Although it is known that the solubility of inorganic nitrogen (e.g., nitrate) is high at all pH levels, the mineralization of organic nitrogen is optimum between a pH of 6.0 and 8.0. In trumao soils, the literature consulted shows that problems with symbiotic fixation activity of nitrogen occur when the pH is below 6.0. Nitrogen fixation improves with applications of calcium carbonate that modifies the pH by making the soil less acid (Urbina, 1982; Russell and Russell, 1967; Thompson, 1965).

Assimilable potassium in an acid soil can be drastically decreased when the soil pH is increased to greater than 7.5. Available potassium may exist in a soil with a pH close to neutral, but it is necessary to fertilize with this element to obtain good yields.

In the pH intervals of a normal soil, the solubility of inorganic sulfur compounds is enough to meet the crop requirement. At a low pH, the quantity of this element is small, while a high pH favors its mineralization from organic forms.

Metallic cations like iron, manganese, copper, and zinc are more soluble if the soil is moderately acid. Iron is more soluble at a pH below 5.0. Manganese, copper, and zinc have greater solubility at a pH level slightly higher than 5.0.

Calcium and magnesium show greater availability at a higher pH unless the soil environment is too alkaline (a pH greater than 8.5). A pH above this level produces a drastic decrease in the quantity of available calcium and magnesium.

Not all crop plants have the same response to soil pH. For example, potatoes, oats, and rye can tolerate medium acidity. Corn, clover, and peas are sensitive to low pH soils. Beets, wheat, barley, and other grains can tolerate a neutral to very weak alkaline pH.

No-till on Santa Lucia de Talermo Farm (an alfisol) close to Chequén, has been established for the last 12 years. It has been observed on this farm that a pH of 5.12 has vigorously stimulated the vinegar weed (*Rumex acetosella* L.) and thistle [*Cirsium vulgare* (Savi) Ten.]. The same situation had been observed on the Montpellier Farm in Mulchén County. In both cases, production had been limited. These weeds have become a serious problem indicated by the lack of vigor of other plants not adapted to a low pH. This points out that **the pH can seriously affect soil fertility and also alter the inventory of plants and weeds.**

Soils with a normal pH can fluctuate between 6.2 and 7.5. At this level, the availability of plant nutrients should not be affected, and the soil microbiology can be developed within normal ranges.

When soils are naturally acid or alkaline caused by factors independent of the action of humans, normally plants and microorganisms can adapt to a low or high pH. The problem arises when the farmer wishes to introduce plants that are not adapted to this environment that may require modification of the pH of the soil.

To reach a desirable level of pH is an easy task, but costly, especially in soils originating from volcanic ash that require greater quantities of calcium carbonate. An analysis of soil is needed to measure pH. The analysis gives a clear indicator of soil acidity or alkalinity; however, environmental factors can cause fluctuations in the pH during the year. These environmental factors include:

- The concentration of carbon dioxide of the soil air (changes in concentration are more sensitive in high pH soils than in low pH soils).
- Presence of any substance that can change the oxidation or reduction potential of the soil.
- The existing vegetation (exerts a major influence on the speed at which bases are lost from the soil).
- The physiologic action of microorganisms in the respiratory processes.
- The availability of nutrients.

All these factors are influenced to a greater or lesser degree by the soils characteristics, so the frequency and quantity of correction processes must be determined after knowing the pH, clay type, aluminum content, precipitation, and the crop to be planted.

Soil acidification can be a normal process with tillage in a climate of abundant precipitation. Bases are leached when high precipitation occurs. Because this process is more man-made than natural, it can be accelerated. This is, in fact, what results from poor management of the soil.

On the other hand, crops extract large quantities of calcium and other base compounds from the soil. The soils can be restored, in the best of cases, to nearly initial conditions if chemical or organic fertilizer is added. At the same time, the traditional practice of plowing leaves the soil bare of vegetation during long periods of the year, often coinciding with periods of abundant rainfall. Selke (1968) indicates that in this way the processes of leaching and, subsequently, the washing

Extensive area affected by vinegar weed (*Rumex acetosella* L.) as a consequence of soil acidification caused by inadequate use or poor management of fertilizers (Source: Montpellier Farm, Mulchén, 1989).

out of exchangeable bases are favored. Studies carried out with lysimeters show that leaching water is rich in basic substances.

Soil pH can be increased in acid soils. The acidity can be neutralized to a degree to satisfy the normal development of plants by:

- Avoiding the use of urea and/or ammonium fertilizer if no inhibitors are used to control the nitrification process.
- Adding fertilizer with calcium.
- Applying a soil conditioner, such as calcium carbonate.

Calcium, besides being an important element in nutrition, can be added and exchanged by organic or mineral colloids of the soil, replacing less positive hydrogen ions; however, excessive quantities of calcium can compete with other cations on the exchange sites, eventually depleting the soil of other nutrients through leaching. Calcium also can precipitate micronutrients such as iron, copper, manganese, and zinc into forms not available to plants.

As already discussed, the acidity of the soil is an equilibrium problem between hydrogen and hydroxyl ions; consequently, soil acidity can be corrected by adding OH^- ions. The best way to correct acidity is to add calcium carbonate. When calcium carbonate is hydrolyzed, Ca^{+2} ions, water, and two hydroxyl ions are liberated. This is expressed in the following chemical equation (Thorup, 1984):

$$CaCO_3 + 2\ H_2O = Ca^{+2} + 2\ OH^- + H_2O + CO_2$$

Table 2–8. Increase in pH level with liming in different soils of southern Chile (Letelier, 1967).

Soil (Series)	Original pH	Liming amount, kilogram of CaO /hectare		
		1000	2000	3000
		Increase in pH		
Granitic (Cauquenes)	5.6	0.5	0.9	1.3
Red (Collipulli)	5.6	0.3	0.6	0.9
Trumaos (Santa Bárbara)	5.8	0.3	0.5	0.6

Soils that have a naturally low pH require constant applications of calcium. Most ammoniacal fertilizers should not be used. The exception is calcium magnesium ammonium nitrate or urea with slow availability of ammonium, (NH_4^+). Increasing the levels of organic matter helps maintain the soil pH by acting as a chemical buffer, resisting acidification. This is probably one reason why it has not been easy to acidify the trumaos soils. These soils are generally rich in organic matter, even when considering that ammoniacal fertilizers have been in use a little more than 30 years. On the other hand, this same phenomenon makes it difficult to increase the pH of these soils, which require greater amounts of calcium carbonate than those required for other soils, including those with a lower organic matter content.

To understand lime quantity or more general lime amendments, the pH, mineralogy of the clay, organic matter percentage, cation-exchange capacity, and percentage of base saturation must be known. The soil should be analyzed by a qualified laboratory to determine how much lime to add.

Lime is commonly referred to as calcium oxide, also called quick lime. Sometimes it is referred to as calcium hydroxide or hydrated lime. The most common name is calcium carbonate. Table 2–8 indicates the quantities of calcium oxide needed to increase the pH of a soil. A granitic soil (alfisol) with additions of 1000 kilograms per hectare (893 pounds per acre) of calcium oxide increases the pH 0.5 units. In contrast, to increase the same 0.5 units of pH in a trumao soil, 2000 kilograms per hectare (1786 pounds per acre) of calcium oxide is required. If the amendment to be used is calcium carbonate, these values would be multiplied by 1.786 (100 kilograms of calcium oxide is equivalent to 178.6 kilograms of calcium carbonate; Donahue et al., 1988).

Not all soils react equally to lime application, nor do all the calcareous amendments have equal liming value. The factors that most influence trumao soils are: organic matter percentage, the type and quantity of clay, and the quantity of extractable aluminum. The greater the level of aluminum, the greater will be the quantity of lime that is required to increase the pH to a specific level (Table 2–9). For this reason, the trumaos soils need greater quantities of calcium carbonate.

Many farmers plow or harrow their soils to incorporate calcium carbonate. In Brazil, in the Río Grande do Sul State, 4 to 5 tonnes per hectare (3572 to 4465 pounds per acre) of calcium carbonate are commonly applied every 3 or 4 years on an eroded soil, which is subsequently harrowed to incorporate it.

Table 2–9. Variation in soil pH in 9 years no-till corn–wheat rotation at Chequén after applying 1500 kilograms per hectare (1340 pounds per acre) calcium carbonate.

Depth, centimeter	Initial pH	Final pH	Variation
0–5	5.77	6.91	1.14
5–10	5.96	6.25	0.29
0–20	5.87	6.41	0.54

The calcium ion can be easily leached if certain conditions are present. To compensate for these losses, lime can be applied on the soil surface. It tends to travel downward to mix uniformly with the soil.

Dr. R.L. Blevins and his associates (1978) established that surface liming is an efficient way to correct soil acidity caused by ammoniacal nitrogen fertilization. Martens and others concur that lime application on the surface in no-tilled crops on a loam soil of Virginia increased corn yields by 20% compared with that on a similar soil under conventional tillage. From this we can deduce that plowing the soil is not necessary to receive the benefits from lime. With surface lime application, equal or better results are obtained (Donahue et al., 1988). **This allows us to conclude that the lime amendment, whatever its origin, does not need to be incorporated in the soil.**

2.10.5. Urea and its Behavior in the Soil

Ammoniacal (NH_4^+) fertilization can decrease the pH. The nitrification processes generate hydrogen ions (H^+) that acidify the soil. This is the principal cause of soil acidification. In Chile, many sources of sodium nitrate and other nitrogen fertilizers exist; however, these fertilizers do not acidify, but actually produce a weak alkaline reaction. For comparative cost reasons, these nitrogen sources have been replaced by urea, which strongly acidifies the soil.

Figure 12 shows what happens when ammoniacal fertilizers are applied. Urea, $CO(NH_2)_2$, hydrolyzes with soil moisture to begin its transformation to ammonium carbonate, $CO_3(NH_4)_2$. This transformation is brought about by the action of an enzyme in the soil called urease. The ammonium carbonate dissociates into ammonium (NH_4^+), carbon dioxide (CO_2) and water (H_2O). In this initial process, in slightly acid, neutral, or moderately alkaline soils, with insufficient moisture; and relatively high temperature, surface application of urea can produce important losses of nitrogen by ammonia (NH_3) volatilization. Ammonia can be phytotoxic.

Ammonium can be absorbed by plants in lower quantities than nitrate; however, the greater part of ammonium is oxidized by action of the nitrifying microorganisms, and is rapidly transformed to nitrate (NO_3^-), water (H_2O), and two hydrogen ions ($2 H^+$). This liberated hydrogen is responsible in a great part for soil acidification as well as the loss of bases or cations, such as calcium (Ca^{+2}), magnesium (Mg^{+2}), potassium (K^+), and sodium (Na^+).

As already explained, the hydrogen ion competes with other cations for space on the soil organic, as well as mineral colloids. These cations pass into the soil solution becoming exposed to important losses by leaching.

Meanwhile, it is not enough to calculate the cost per nitrogen unit without valuing its efficiency and benefit to the plant and soil. **Applying fertilizers of low cost per unit may not be beneficial if its behavior in the soil and plants is not considered.**

Fertilization in bands next to the seed is generally beneficial. The concentration of nitrogen fertilizer, like ammonium and ammonia, can influence crop growth negatively under certain conditions of moisture and temperature and be toxic to plant germination and seedling growth. The application of 150 kilograms per hectare (134 pounds per acre) of urea on winter oats planted May was strongly affected in Chequén. This was probably caused by the release of ammonia gases during the process of urea transformation. In addition to this problem, nitrogen loss by volatilization and a lower pH in the rhizosphere occurred from the use of urea. The problems worsened because of inadequate moisture and temperatures higher than 15°C (60°F).

R.D. Meyer, R.A. Olson, and H.F. Rhoades, professors in the Department of Agronomy at the University of Nebraska, say that urea must be incorporated into the soil to assure its maximum efficiency. Otherwise, rain or irrigation should occur immediately after the surface application (Thorup, 1984). Lewis B. Nelson, editor–consultant of the *Farm Chemicals* magazine, said that

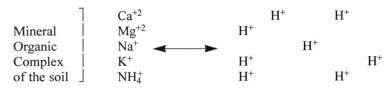

UREA

45% Nitrogen

Reaction of Urea in the Soil

$$\uparrow$$
$$CO(NH_2)_2 + \frac{H_2O}{urease} \rightarrow CO_3(NH_4)_2 \rightarrow NH_4^+ + CO_2 + H_2O$$

Reaction of Ammonium in the Soil

$$NH_4^+ + 2O_2 \rightarrow NO_3^- + H_2O + 2\,H^+$$
$$\uparrow$$

Nitrifying Microorganisms

Reaction of pH in the Soil

$$2\,H^+ \ + \ Ca^{+2}\,(Clay) \ \rightarrow \ Ca^{+2} \ + \ H^+\,(Clay)$$
$$\downarrow$$

Leached   Acidification
of the Soil

	Ca^{+2}		H^+	H^+
Mineral	Mg^{+2}	H^+		
Organic	Na^+ ⟷		H^+	
Complex	K^+	H^+		H^+
of the soil	NH_4^+	H^+	H^+	

Fig. 12. Biochemical transformation of urea in the soil.

high losses by ammonia volatilization (NH_3) can occur during the transformation of urea if it is applied on the surface in acid or alkaline soils. He added that, as a general rule, urea should not be applied on the surface of the soil without immediate incorporation (Thorup, 1984).

Because of the problems urea causes when broadcast on the surface without incorporation, specialists recommend a substitute of ammonium nitrate or preferably calcium magnesium ammonium nitrate (Thorup, 1984). Thus fertilizers that contribute to soil acidification are replaced by those that possess neutral or weak alkaline reaction.

2.10.5.1. Corn, Urea, and pH at Chequén

A tendency to lower pH was observed at Chequén during the first 7 years of no-till management. Table 2–10 shows that in no-till the pH decreased from 6.14 to 5.77. This gradual alteration of the pH signified a serious problem in the nutrition of corn fertilized with ammonium phosphate and urea, which are both acid forming fertilizers.

From the first plantings of no-till corn, a lack of vigor of 1-month-old plants was observed. The crop residue on the surface lowered soil temperature in spring and made it difficult to manage physical aspects as basic as temperature. For this reason, planting was done 1 week later; however, the same lack of vigor resulted. After 7 years, the deficient vigor in plants was strongly detected, indicating that factors other than the lack of temperature, as so often observed by many specialists, were having an effect. At the end of November 1986, 160 millimeters (6.4 inches) of

Table 2–10. Changes in soil pH due to ammonium fertilizers (urea and ammonium phosphate) in a 7 year, no-till corn–wheat rotation at Chequén.

Depth, centimeter	Initial pH	Final pH	Differential
0–5	6.14	5.77	−0.37
5–10	6.03	5.96	−0.07
10–20	5.86	5.87	+0.01

rain fell in 48 hours, which was much greater than normally expected. The problem worsened with intensive chlorosis and lack of vigor in the 1-month-old corn plants. The ammoniacal fertilizer applied (urea and ammonium phosphate) increased soil acidity to a pH of 4.8 in the germination zone adjacent to seed.

A loss of calcium and magnesium in the soil resulted from leaching, which were then replaced by hydrogen ions on the soil colloids. Amberger pointed out that as a general rule, ammoniacal nutrition by general rule, tends to decrease assimilation and conservation of calcium and magnesium in the plant (Romheld, 1986). Other investigators point out that if the soil pH is kept low, one can expect a decrease in calcium and magnesium levels and, at the same time, an increase in aluminum and manganese in the soil solution (Blevins et al., 1985).

Table 2–10 shows that a lower pH is produced in the surface horizon where ammonium fertilizers were applied. Therefore, the calcareous amendments should be applied on the soil surface and not incorporated because the increase in soil acidity occurred only at the surface.

As already explained, corn plants require a pH close to neutral to reach their maximum productive potential. To clearly determine which deficiency was affecting the corn, a foliar analysis was made. The results are shown in Table 2–11. The analysis detected 0.22% calcium, which is classified as very low. The normal calcium level is between 2 and 4%. Magnesium was close to the critical level of 0.26%. Normal values are 0.26 to 0.60%. To conclude, the corn had both a calcium and magnesium deficiency.

The truth is, we did not pay enough attention to secondary nutritional elements, such as calcium and magnesium. This bad experience reduced the corn yield by 3 tonnes per hectare (48 bushels per acre) in the no-till corn field affected by soil acidity. All this happened because of the high rainfall in the spring. Without the foliar analysis it probably would have taken a longer time to understand the real problem. Today, I do not use any urea or ammonium phosphate at Chequén Farm, and the corn yields increase every year. We are paying more attention to the correct pH and using the proper fertilizer.

2.11. CATION-EXCHANGE CAPACITY

The cation-exchange capacity is important because it forms the fundamental mechanism involved in plants obtaining nutrients from the soil. Understanding this mechanism of cation inter-

Table 2–11. Foliar analysis in corn plants affected in development by the application of 300 kilograms of urea per hectare (268 pounds per acre) applied with the planter and placed in bands on either side of the seed.

Element	%	Level
Nitrogen	4.09	Normal
Phosphorus	0.31	Normal
Potassium	3.75	High
Calcium	0.22	Low
Magnesium	0.26	Minimum

Chlorosis in no-till corn affected by soil acidification resulting from the application of ammoniacal fertilizers (1986).

Corn crop after the correction of soil acidity. The plants appear vigorous and will be more productive (1988).

change is vital for farmers who manage and fertilize their own soil while striving to accomplish conservation and soil improvement while increasing production.

The concepts that follow are based on my point of view and experiences in constantly studying the phenomenon occurring at Chequén as well as the results of different chemical–physical–biological soil analyses that have been made on a regularly basis. It is necessary to explain some elemental aspects so that nutritional processes of the plant are better understood.

First, an ion is defined as an atom or particle that can lose or gain one or more electrons. It can acquire an electrical charge. When an ion loses an electron, the charge is positive and the ion is called a cation. When an electron is gained, the charge is negative and the ion is called an anion. As we all know, the electrical charges of different polarity are attracted to each other in a very strong way, and similar charges are repelled. This elemental principle is important in the nutrition of plants.

Ion exchange is the reversible process by which solid particles of the soil, called *exchangers*, are able to adsorb ions on their surface from the soil solution. The soil water has many nutrients in the soluble form and in equilibrium with the colloids. These ions are the basic source of nutrition for plants. This exchange of ions must be accompanied simultaneously by a balance of cations (with positive electrical charges) that adhere to the soil surface (with a negative charge) with cations that go into solution. This establishes an equilibrium and provides available nutrients for plants.

The chemical and physical processes related to ionic exchange include the structure of the soil minerals, nutrient absorption by the plants, clay development and transformation, and the leaching of soluble elements. This is a physical–chemical phenomenon caused by the electrical charges of the soil colloidal particles, which can attract cations from the soil solution. The soil must be moist before and during the exchange process for the exchange to occur and for a rapid equilibrium exchange to be reached (Thompson, 1965).

The colloids that have a higher cation-exchange capacity are generally those that have a greater specific surface area. To understand the term specific surface, I refer to the classic example cited in most literature. Consider a cube 1 meter high by 1 meter wide by 1 meter long (1 cubic meter). The specific surface of this cube is the sum of the surfaces of each face. Because the cube has six faces, it has a surface area of 6 square meters. If one cuts the cube in half along both the vertical and horizontal axis at each of its faces, the result is eight small cubes each having a volume equal to 0.125 cubic meters and a surface area of 1.5 square meters. Since there are now eight cubes, the surface area of the original cube has increased from 6 square meters to 12 square meters. As the cubes are subdivided into smaller cubes, the initial volume of the original cube is maintained, but the surface area increases.

Transferring this same concept to the soil, a volume of particles the size of sand can be estimated. If this surface area is compared with the same volume made up of particles the size of clay (which are much smaller than sand particles), a much higher surface area will be found. Mineral and organic colloids have a high specific surface area. Colloids from organic matter have a specific surface area between 500 and 800 square meters per gram. The colloids constituted by montmorillonitic clays have a specific surface area of 800 square meters per gram. Illitic clays have 50 to 100 square meters of specific surface area per gram (Robinson, 1967; Bear, 1963).

This cation-exchange capacity phenomenon apparently works incredibly well since clays possess different agronomic qualities. The much finer clay produces a better opportunity for good soil fertility needed in agricultural production. The same thing happens with the contribution from organic matter, which enriches the soil with organic colloids. They can significantly improve the cation-exchange capacity and maintain the fertility of the soil.

Table 2–12. Percentages of optimum levels of cations on the soil cation-exchange capacity for good production (Cargill Chile, Ltd., 1988).

Cations	Percentage base saturation
Calcium	60–70
Magnesium	10–20
Hydrogen	10–15
Potassium	2–5
Others†	2–4

† Includes iron, manganese, cooper, zinc, and sodium.

The calcium cation is an important contributor to the cation-exchange capacity. For this reason it must be maintained within the ranges shown in Table 2–12. Magnesium, potassium, sodium, and ammonium also are important. These cations can be absorbed by the soil colloids.

The soil's buffering capacity is manifest when a supply of chemical fertilizer is in the soil that increases the concentration of the principal nutrient ions in the soil solution. Under these conditions, the soil adsorbs part of the nutrient concentration on the surface of the particles, but at the same time, releases other cations to the solution, trying to establish an equilibrium between the soil particles and the soil solution. This impedes buildup of concentrations in the soil solution to toxic levels. Equally, when the solution becomes nutrient-poor by plant absorption, the soil particles free ions from their surface. In this way, the elements considered essential for growth are provided to the plants.

Cations can be grouped into two forms:

(1) The dominant basic cations of exchanges: calcium, magnesium, potassium, and sodium.
(2) The dominant acidifying cations of exchange: aluminum, iron, manganese, hydrogen, and ammonium.

The sum of both groups characterize the maximum adsorption capacity of cations by the soil particle surface. This is termed the cation-exchange capacity. See Fig. 13.

From the point of view of the farmer, the soil exchange complex fulfills an important role in plant nutrition by reserving the indispensable cations. If this form of storage were not present in the soil, rains would leach the cations and leave behind a completely inert soil (Gaucher, 1971). This is one reason why sandy soils have lower fertility than clayey soils. Probably the only way to improve the nutrient retention in sandy soils is to elevate the organic matter levels and, at the same time, avoid using tillage.

When we discuss soil particles we are referring to the small soil particles less than 0.002 mm, specifically the colloids. Humus resulting from mineralization of organic matter also is this size. For practical purposes reference will be made to the *colloidal complex*, which includes both clay and organic matter.

Table 2–13 shows the values of cation-exchange capacity for different types of clays including organic matter. Clays with the greatest cation-exchange capacity are montmorillonites, vermiculites, and halloysites (Garavito, 1979).

Mineral colloids have an amphoteric character; that is, they can absorb anions as well as cations. This is caused by a variable charge that is dependent on the pH of the soil solution and the presence of aluminum and iron oxides. The neutralization of the charge of these oxides produces a determined pH, known as the *isoelectric point* or neutral point, which is found in the acid soil (pH around 4.0). When these oxides are found at a pH above the isoelectric point, they are negatively charged. At a pH below the isoelectric point, they are positively charged. Under spe-

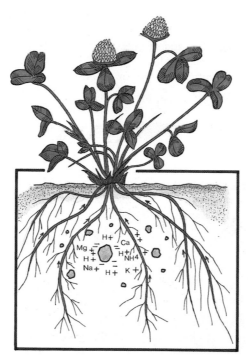

Fig. 13. Cation-exchange capacity and plant nutrition.

cific pH conditions the colloidal complex can adsorb cations and, in others cases, anions. It is evident then that the soil pH is important in the anionic complex and, when the pH is lower than the isoelectric point, the adsorption of anions increases and adsorption of cations decreases.

Clays, such as montmorillonites, vermiculites, and others that have a silicon-aluminum 2:1 relation, have negative charges that are not affected by pH. No equilibrium or isoelectric point exists. This indicates that in these clays, negative charges always predominate. On the other hand,

Table 2–13. Cation-exchange capacity of the principal soil colloids (Garavito, 1979)

Material	Cation-exchange capacity
	milliequivalents per 100 grams of soil
Iron and aluminum oxide	2
Iron and aluminum amorphs	2
Kaolinite	3–15
Sepiolite/Atapulgite	3–15
Halloysite 2 H_2O	5–10
Illite	10–40
Chlorite	10–40
Allophane	25–50
Halloysite 4 H_2O	40–50
Smectites (montmorillonites)	80–150
Vermiculites	100–150
Peat moss	100–150
Organic material†	150–250
Soils	0–50

† Some organic compounds have values up to 500 milliequivalents per 100 grams of soil.

the clays with a 1:1 relation, such as kaolinite (Chequén soils) and allophanic soils (trumaos), present negative or positive charges, depending on the pH of the solution (Garavito, 1979). Soils of 1:1 relation are dominant across Chile and, the pH of these soils should be maintained as close as possible to neutral. This can be done by adding lime and increasing organic matter.

The following anions are normally in the soil solution:

Phosphate PO_4^{-3}, HPO_4^{-2}, $H_2PO_4^{-}$
Sulphate SO_4^{-2}
Carbonate CO_3^{-}
Nitrate NO_3^{-}
Chloride Cl^{-}

Intense rains cause heavy losses of these anions except in medium to fine textured soils that have little vegetation (winter season). On the other hand, anion adsorption by soil colloids that have amphoteric characteristics depends on the following factors:

(a) Nature of the adsorbent
(b) Surface area of the adsorbent
(c) Adsorbed anions
(d) Soil temperature
(e) Concentration of anions in the liquid phase

Because of the many intervening factors, an anion exchange capacity cannot be compared with a cation-exchange capacity. Distinct anions are adsorbed by different mechanisms other than an electrical charge on the soil. It is possible to determine the anion-exchange capacity of a certain anion only under specific conditions.

In trumao soils, the anion-fixation capacity, especially for phosphates, by the active presence of allophane and related materials is of such high magnitude that large doses of monocalcic phosphates must be applied relative to the crop requirements (Zunino, 1983).

Alfisols (haploxeralfs) have a low cation-exchange capacity because of the dominant type of clay in the colloidal complex (kaolinite). The clay has a 1:1 relation; that is, the silicon and aluminum are present in equal parts; however, in predominantly clay soils, such as vermiculite or montmorillonite, the relation is 2:1, which signifies that silicon is dominant over aluminum. This type clay provides the soils with better chemical properties, especially in base cation exchange properties.

The edaphologic characteristic can be improved in all soils that have low cation-exchange capacity by increasing the organic matter content through no-tillage. This is shown in Table 2–13.

Analysis of the organic matter content of distinct soil profiles of Chequén soils show the positive effect of organic matter especially in the first 5 centimeters (2 inches) of the soil surface layer (Fig. 14). The analysis shows the changes with cation-exchange capacity under different types of management.

Table 2–14 shows the relationship of organic matter and cation-exchange capacity under three management systems:

• conventional, that includes plowing, harrowing, disking, and other tillage practices
• pasture
• no-till

One can see in the first system that the cation-exchange capacity is low because the organic matter level is low. In the management of permanent pasture (more than 15 years) where the soil has not been disturbed and fertilizer has been added only to maintain productive pasture, one

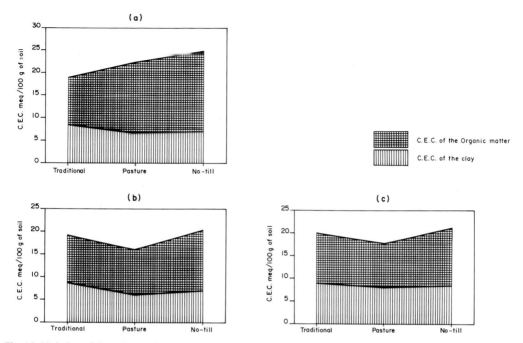

Fig. 14. Variation of the cation-exchange capacity in relation to the organic matter content under three management systems in Chequén soils: (a) 0 to 5 centimeters; (b) 5 to 10 centimeters; and (c) 10 to 20 centimeters.

sees a notable improvement in both the cation-exchange capacity and organic matter levels. In this case the supply of residue is limited and only dry material, such as straw or hay, is removed. No-till results in the highest levels of organic matter and cation-exchange capacity in the top 20 centimeters (8 inches). The top 20 centimeters of the soil was the most active horizon.

The soil cation-exchange capacity corresponds to the sum of the cation-exchange capacity of the organic colloids plus that of the mineral colloids. Total cation-exchange capacity is determined normally in a soil laboratory and can be expressed as follows:

$$\text{Total CEC} = \frac{\% \text{ orgainc matter} \times CEC_{om}}{100} + \frac{\% \text{ clay} \times CEC_{clay}}{100}$$

Table 2–14. Relation between the organic matter content and cation-exchange capacity of soil profiles at Chequén, under three systems of management.

Management system	Depth	Organic matter	Cation-exchange capacity
	centimeter (inches)	%	milliequivalents per 100 g of soil
Conventional	0–5 (0–2)	1.42	11
5 years wheat	5–10 (2–4)	1.24	11
	10–20 (4–8)	1.00	11
Pasture	0–5 (0–2)	4.56	16
(+15 years)	5–10 (2–4)	1.92	10
	10–20 (4–8)	1.14	10
No-till	0–5 (0–2)	5.32	18
7 years	5–10 (2–4)	2.84	13
Wheat–corn	10–20 (4–8)	2.24	13

Table 2–15. Variation of cation-exchange capacity in relation to soil colloids under three management systems at Chequén.

Management system	Depth	Cation-exchange total	Cation-exchange capacity organic matter†	Cation-exchange capacity clay‡
	centimeter (inch)	——————— milliequivalents per 100 grams of soil ———————		
Conventional	0–5 (0–2)	11	2.83	8.17
5 years wheat	5–10 (2–4)	11	2.47	8.53
	10–20 (4–8)	11	2.00	9.00
Permanent	0–5 (0–2)	16	9.11	6.89
Pasture	5–10 (2–4)	10	3.97	6.03
(+ 15 years)	10–20 (4–8)	10	2.27	7.73
No-till	0–5 (0–2)	18	10.63	7.37
(7 years)	5–10 (2–4)	13	5.67	7.33
Wheat–corn	10–20 (4–8)	13	4.47	8.53

† Cation-exchange capacity due to organic matter.
‡ Cation-exchange capacity due to clay.

From this estimated relationship, one can recognize two important aspects:

1. Assuming that the cation-exchange capacity of the organic matter (CEC_{om}) is 200 milliequivalents per 100 grams as an average value from Table 2–13, one can estimate the cation-exchange capacity from the clay (CEC_{clay}). Comparing the resulting value with those that are shown in Table 2–15, one can evaluate which clay is predominant in the mineral colloidal complex.
2. Estimating the type of predominant clay in the mineral colloidal complex, one can evaluate the cation-exchange capacity due to organic matter.

Applying this concept to that which occurred at Chequén (Table 2–15 and Fig. 14), one appreciates that because of the type of clay, the cation-exchange capacity values are practically constant under the three management systems, given similar soils with similar clay content. The significant increases of the total cation-exchange capacity in pasture and no-till, with respect to the conventional system, is caused by the increase in soil organic matter. All this reaffirms one more time the need to conserve the residue over the soil to obtain such extraordinary benefits.

Another factor that affects the soil fertility and has a direct relationship with the cation-exchange capacity is the percentage of base saturation (%BS). This percentage is the total cation-exchange capacity occupied by each one of the principal cations in plant nutrition. The optimum percentages are shown in Table 2–12. This table explains why acid soils with a large quantity of hydrogen ions replacing calcium, magnesium, and other soil cations, have low soil fertility. This phenomenon occurs each time acidifying fertilizers, such as urea or ammoniacal fertilizers, are applied. Urea, upon ammonifying and nitrifying, can generate hydrogen ions, which can replace useful cations (mass conservation law). The cations displaced from the soil colloid enter the soil solution and are exposed to leaching from excess rain or irrigation (Fig. 12).

Aluminum can interfere in the soil cation-exchange capacity upon occupying spaces that otherwise could be filled by useful cations. It is generally abundant in granitic soils of the coastal foothills and even more in the trumao soils of the Andean foothills.

In regions with low rainfall and well-drained soils, Ca^{+2} and Mg^{+2} are adsorbed first on the cation exchange sites, followed by K^+ and Na^+, and finally H^+. When drainage conditions of these soils are not good, salts accumulate at the surface, and Na^+ can be adsorbed in equal or greater quantities than Ca^{+2}. The hydrogen ion occupies a preferential place in humid regions and of less importance in low-rainfall regions.

Table 2–16. Variation of bases on the exchange complex in soils of Chequén under three management systems.

Management system	Depth centimeter (inch)	Cation-exchange capacity	Calcium	Magnesium	Potassium	Sodium	Sum	% Base Satration
Conventional	0–5 (0–2)	11	4.75	2.02	0.49	0.25	7.51	68.27
5 years wheat	5–10 (2–4)	11	5.50	2.14	0.47	0.27	8.38	76.18
	10–20 (4–8)	11	4.88	2.02	0.45	0.29	7.64	69.45
Pasture	0–5 (0–2)	16	9.13	2.59	0.56	0.30	12.58	78.63
(+15 years)	5–10 (2–4)	10	5.50	2.06	0.52	0.28	8.36	83.60
	10–20 (4–8)	10	4.88	1.93	0.56	0.22	7.59	75.90
No-till	0–5 (0–2)	18	12.00	2.43	0.83	0.32	15.58	86.56
7 years	5–10 (2–4)	13	7.00	1.67	0.61	0.24	9.52	73.23
Wheat–corn	10–20 (4–8)	13	7.00	2.06	0.47	0.25	9.78	75.23

Table 2–16 shows the variation of the bases on the exchange complex that have been observed in the soils of Chequén. The three management systems already mentioned are compared.

One observes that the base saturation percentage is generally high, more than 50%. This includes the soils under traditional management with predominance of calcium (Ca^{+2}) on the exchange complex. The sequence of importance of the cations starting with Ca^{+2} being the most important is: Ca^{+2}, Mg^{+2}, K^+, and Na^+.

In this way, the exchange complex is saturated with more than 40% Ca^{+2} in the first 5 centimeters (2 inches) of soil managed in traditional form. This value increases to more than 66% in no-till, as shown in Table 2–17.

The results speak for themselves as to the benefit of no-till. This is especially true if the values of saturation percentage of each cation are compared with the optimum levels for good fertility and plant nutrition as shown in Table 2–12.

The results obtained over the years are important. An elevated cation-exchange capacity established with a high saturation percentage of predominantly calcium represents good fertility as well as good soil physical characteristics. The results at Chequén indicated that the **decline of calcium on the exchange complex and the acidification of the soil are indicators of the degradation of cultivated soils, besides being signs of poor agronomic practices.**

Table 2–17. Variation of percentage of base saturation in soils of Chequén under three management systems.

Management System	Depth centimeter (inch)	Calcium	Magnesium	Potassium	Sodium	Hydrogen†	Total
Conventional	0–5 (0–2)	43.18	18.36	4.45	2.27	31.73	100
5 years wheat	5–10 (2–4)	50.00	19.45	4.27	2.46	23.82	100
	10–20 (4–8)	44.36	18.36	4.09	2.64	30.55	100
Pasture	0–5 (0–2)	57.06	16.19	3.50	1.88	21.37	100
(+15 years)	5–10 (2–4)	55.00	20.60	5.20	2.80	16.40	100
	10–20 (4–8)	48.80	19.30	5.60	2.20	24.10	100
No-till	0–5 (0–2)	66.67	13.50	4.61	1.78	13.44	100
7 years	5–10 (2–4)	53.84	12.85	4.69	1.85	26.77	100
Wheat–corn	10–20 (4–8)	53.84	15.85	3.62	1.92	24.77	100

† Other cations besides H^+ included.

Table 2–18. Percentage of decomposition of organic substrates marked with C^{14} in allophanic and nonallophanic soils (Zunino, 1983).

	Evolution of carbon from substrate (% C^{14} eliminated in 5 months)			
	Nonallophanic soils		Allophanic soils	
	Concentration of added allophane %			
Substrate	0	5	16	
Glucose	78	66	58	56
Cellulose	75	56	50	34
Wheat straw	60	41	36	33
Cells of Hendersonula toruloidea	32	21	19	17
Cells of Mucor r.	56	48	35	39

2.12. TRUMAO SOILS (ANDEPTS) AND ORGANIC MATTER

One of the most important characteristics of trumao soils is the high levels of organic matter that they possess. Even if these soils are poorly managed using tillage implements for winter or spring crops and burning crop residue, a rapid decline has not been observed in the organic matter levels. This still permits an excellent absorption of rainwater. Organic matter levels are approximately 8 to 20% (Bernardi et al., 1973; Zunino, 1982). Nevertheless, if these actual systems of cultivation are continued, one cannot hope or expect that these conditions will be maintained.

The large quantity of aluminum in trumao soils (commonly 1500 parts per million) causes the processes of biodegradation of organic matter to be different from those in other soils. Allophanic soils of high aluminum content interfere with the decomposition of organic matter, decreasing the rate of CO_2 evolution. These soils have special physical, chemical, and biological properties that generate a soil microbiology dominated by actinomycetes and fungi. These microorganisms have a behavior different from that of normal soils (Zunino, 1983).

Trumao soils tend to accumulate humus, which slows the processes of mineralization of organic matter. The aluminum is considered as the essential constituent for the stabilization of organic matter against biotic degradation and leaching (Besoain and Sepulveda, 1985).

Dr. Hugo Zunino developed the information in Table 2–18 to better explain how allophane interferes in the degradation of organic matter and the scientific justification of its high levels in these soils under cultivation.

Table 2–18 clearly shows the effect of allophane on the decrease of carbon dioxide evolution of the added organic material and in particular in the different substrates subjected to a concentration of 5 and 16% allophane content. As the quantity of concentrate increased, the evolution of radioactive carbon ($^{14}CO_2$) decreased, which could indicate a possible decrease in soil respiration activity. This inhibition of the metabolism of the microorganisms responsible for the decomposition of organic residue could be the most important cause of the high content of organic matter in trumao soils.

I will try to explain this phenomenon through the chemical behavior of aluminum expressed as the Al^{+3} ion, which is hydrated in aqueous solutions. The hydrolysis of the aluminum salts, from the hydrated aluminum ions, lose protons (H^+) forming successive complex hydroxides. The final neutral complex loses water, becoming transformed in insoluble $Al(OH)_3$ hydroxide. This can be explained through the following equations:

$$Al\,(H_2O)_6^{+3} \quad \rightleftharpoons \quad [Al\,(H_2O)_5\,OH]^{+2} + H^+$$

$$Al\,(H_2O)_5\,OH^{+2} \quad \rightleftharpoons \quad [Al\,(H_2O)_4\,(OH)_2]^{\,+} + H^+$$

$$Al\,(H_2O)_4\,(OH)_2^{+} \quad \rightleftharpoons \quad [Al\,(H_2O)_3\,(OH)_3] + H^+$$

$$Al\,(H_2O)_3\,(OH)_3 \quad \rightleftharpoons \quad Al\,(OH)_3 + H_2O + H^+$$

As one can observe, the concentration of the hydrogen ions produced in this way produces additional acidity, and the final stage of hydrolysis produces the precipitation of hydroxide $Al(OH)_3$, but only if the concentration of the hydrogen ion in the dissolution decreases (Pauling, 1958).

The high concentration of hydrogen ions implies a low pH, which would explain the origin of the acidity of the soils derived from volcanic ash.

No-till contributes efficiently to a better absorption of such chemical elements as phosphorus in trumao soils. This permits the recycling of the residue and introduces *clean* organic matter not contaminated by aluminum. Phosphorus, together with nitrogen and sulfur, forms a part of organic combinations in residue (Zunino, 1983; Muñoz, 1990). A large part of the phosphorus in trumao soils with no-till and their residue compared with tilled soils and burned residue, comes from the mineralization of these organic combinations. It is important to point out that **even the minimum disturbance of soil, after years of no-till, can expose and reactivate the phosphorus-fixation sites.**

2.13. Fertilization with the Planter

2.13.1. Nitrogen

The application of nitrogen with the planter is vitally important especially in cereal crops. Sufficient nitrogen also is needed during the early development of plants. An adequate initial fertilizer application to the crops accelerates their growth, gets them in condition to compete well with weeds, and increases yields. Nitrogen applied this way can provide conditions of greater initial vigor, shortening the seedling period. It also helps to protect the plant from problems of low temperatures, moisture excesses, and crop pests.

The quantity and type of nitrogen to apply with the planter depends on the soil class, acidity, crop planted, climate, and planting date. These factors have a direct relation with the possible losses of nitrogen by leaching and denitrification.

When planting winter wheat on Chequéns clay soils, we traditionally applied 40% of the total nitrogen next to the seed in the form of urea. Later, because of problems of acidification, urea was replaced by sodium nitrate. Since 1990, only calcium magnesium ammonium nitrate (Nitromag, Cargill) has been used at planting and tillering of the crop. This fertilizer presents exceptional qualities in soils with relatively low pH and does not acidify because the granule is covered with dolomitic limestone. This improves the availability of calcium and magnesium for the plants. The assimilation of this fertilizer by plants is primarily in the nitrate fraction, using the ammonia at a later growth period. The main reason for this is that the ammonium (NH_4^+) fraction is retained by the soil colloidal site, which the plant does not use, and is transformed into nitrate by biological oxidation. In this way, the plant has adequate nitrogen fertilizer available at the time it is needed, minimizing leaching losses. Table 2–19 shows the principal characteristics of nitrogen fertilizers and their compounds. The fertilizers shown are available on the national market.

Table 2–19. Principal nitrogen fertilizers.

Type	Fertilizer	Nitrogen	Phosphorus	Potassium	Calcium	Magnesium	Sulfur	Sodium	Form of Application
Nitrate	Sodium nitrate	16	-	-	-	-	-	16	Incorporate or on surface. Produces weak alkalinization.
	Calcium nitrate	16	-	-	26	-	-	-	
	Potassium nitrate	13	-	44	-	-	-	-	
Nitrate ammonium	Ammonium nitrite	33	-	-	-	-	-	-	Placed with the seed in cool, moist weather; incorporate or on the surface
	Calcium magnesium Ammonium nitrate	27	-	-	7	5	-	-	
Ammonium	Monoammonium phosphate	10	50	-	-	-	7	-	Reacts in neutral or slightly acid soils.
	Diammonium	18	46	-	-	-	-	-	Needs to be incorporated
	Ammonium sulfate	21	-	-	-	-	24	-	Reacts in neutral or slightly alkaline soils
	Urea	45	-	-	-	-	-	-	Requires a process of transformation to ammonium carbonate in the soil, then to nitrogen.

The losses of nitrate nitrogen in coarse-textured soils (sandy and trumaos) can be high during heavy rains. For this reason it is advisable to spread nitrogen in low doses but with more applications.

In no-till, nitrogen tends to immobilize in the residue, and the quantities added prior to planting are not always adequate for early crop growth. During decomposition of the previous crop roots, some of the fertilizer nitrogen that was applied with the planter or broadcast is used; however, when these roots decompose they release the nitrogen for plant uptake.

2.13.2. Phosphorus

Phosphorus fixation is a common phenomenon in soils of southern Chile, especially in the trumaos of the Andean foothills. The large demand for this element results from the gradual loss by erosion of the original organic profile and to a low pH caused by poor fertilization. In this case the high aluminum content of the soil must be considered because it is responsible, in great part, for the fixation of phosphorus.

Phosphorus is fixed in the trumao soils by chemical precipitation with aluminum and iron. The entire phosphorus requirement of the plant can be applied with the seed. To favor plant nutrition and avoid major fixation losses, the application should be confined to a narrow band next to or with the seed.

2.13.3. Soil Aluminum and Phosphate Solubility

The presence of extractable or free aluminum (Al^{+3}) in cultivated soils causes a high fixation of the soluble phosphates. This is especially true for monocalcium, such as ammonium phosphate, triple superphosphate, and normal superphosphate. Ammonium phosphate and triple superphosphate are 98% water soluble, while the normal superphosphate is 32% water soluble. For this reason, normal superphosphate presents less potential phosphorus fixation in acid soils. The remaining fraction is only soluble to organic acids generated by the soil microorganisms.

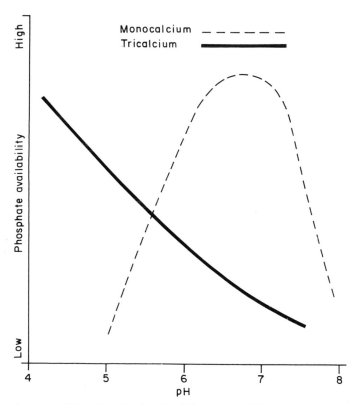

Fig. 15. Availability of mono- and tricalcium phosphate in relation to soil acidity.

With respect to this, Dr. Anton Amberger of the University of München, in his visit to Chile in 1989, said that the fixation of the monocalcium phosphates applied in acid soils of the region can happen in just a few minutes. He recommended using a soil analysis to determine the pH and available phosphorus, and, according to the results, applying phosphoric rock. This is an efficient and economic alternative to improve the availability of phosphorus and the pH of these soils.

To better understand the importance that soil pH has in the availability of phosphorus, consider that in a soil with a pH drop of one unit from neutral, the availability of phosphorus is 10 times less than that in a neutral pH soil. If the pH drops to 5.0, the decrease is 100 times less. The phosphorus availability continues decreasing in a logarithmic reciprocal relationship (Thomas, 1986). This indicates the need to ensure that the pH of the soil does not decrease from normal levels. In acid soils, apply tricalcium phosphates to decrease the loss by fixation. As shown in Fig. 15, the range of major availability of monocalcium phosphate occurs between pH 6.0 and 7.5. This indicates the area where monocalcium phosphates might be used.

It is estimated in Chile more than 90% of the phosphate fertilization on trumao soils comes from the monocalcium phosphates. This is a big mistake in these soils and makes producing cash crops extremely difficult.

In acid soils, the fixation processes of monocalcium phosphates is mainly because of the high levels of extractable aluminum. These negative processes can be stimulated by:

- Continuous cropping, which can cause soil, organic matter, and nutrient loss by erosion, loss of bases by leaching, and decrease in pH.

- Loss of crop residue by burning. Soils with a high content of organic matter offer a major resistance to the change of the natural soil pH (buffering capacity).
- Tilling a soil that has been fertilized with monocalcium phosphates and then fertilizing the newly exposed soil, starts again the processes of phosphorus fixation.

No-till permits with time the phosphate fertilizer to become fixed in one horizon, concentrating it and assuring the availability for the crops in a more available form. This is done by not tilling the soil in successive croppings. Although phosphate does not have great physical movement in the soil, we have observed that under no-till, a high phosphate concentration is available for plant nutrition in a layer from 5 to 10 centimeters (2 to 4 inches) deep. This indicates that phosphate has biological movement through mesofauna and the plant root system (see Table 4–3).

2.13.4. Phosphate Rock and No-till

It has been mentioned that the monocalcium or water-soluble phosphates are easily fixed by a process that can become irreversible. Farmers that work with trumao soils must increase the phosphorus amounts to counterattack the strong adsorption capacity of the sesquioxides of aluminum and iron present in the soil. The benefit of the additional phosphorus is to obtain adequate yields. If a specific crop requires 40 to 60 units of P_2O_5 per hectare (36 to 54 pounds per acre) in trumao soils, between 160 and 200 units (143 to 179 pounds per acre) must be applied. This indicates a serious adsorption loss or potential deficiency in the phosphorus availability to the plants.

The use of tricalcium phosphate or phosphate rock are great alternatives in soils with high phosphorus fixation that are affected by elevated levels of extractable aluminum and iron and high acidity.

Phosphate rock can be divided into three groups: igneous, metamorphic, and sedimentary, all of which belong to apatite mineral phosphates. The first two are normally marketed by the same mineral producing nations. The sedimentary phosphate rock constitute the principal source of commercial phosphate and are the basis for the fabrication of monocalcium or water-soluble phosphates. These apatites can be very different. They are found in the form of oxide, hydroxide, carbonate, and fluoroapatite because of different chemical compositions occurring in distinct geologic periods. Most come from the miocene (Tertiary era) through the Precambrian era (Hill, 1976).

Considering the quality and reactivity of the phosphate rock, these can be a good source of phosphate for the plants in acid soils, especially when the index of relative solubility is more than 60%. This index, an arbitrary value, has been established comparing phosphoric rocks with those of North Carolina (80 to 100%), that is considered one of the best natural deposits in the world. This phosphate rock has sufficient reactivity to be applied directly to the soil without any chemical treatment. Other phosphoric rock deposits with a relative solubility index of more than 60% are in Israel, Tunisia, Peru, and others parts of the world.

The relative solubility of the different deposits of phosphoric rock are shown in Table 2–20.

If the phosphate rock has a relative solubility index of more than 60%, it is adequate for direct fertilization. This can be of great importance in the fertilization of acid soils on which vegetative cover exists (Fig. 15).

Figure 15 indicates that the efficiency of the tricalcium phosphates in soils with low pH causes greater use of phosphorus by plants. Nevertheless, when plants or residue are not present to absorb available phosphorus, fixation of this labile phosphate can occur on the soil colloids. On the other hand, the use of monocalcic phosphates decrease as the pH of the soil solution decreases or increases from pH 7.0. The tricalcium phosphates do not react with extractable aluminum

Table 2–20. Relative comparisons of different origin and solubility of phosphorous.

Origin	20%	40%	60%	80%	100%†
Slag (local)	XX				
Western USA	XXXXXXXX				
Tennessee		XXXXXXXXXXXXX	XX		
Morocco		XXXXXXXX	XXXX		
Florida		XXXXXXXXX	XXXXXXXXXXX		
Israel		XX	XXXXXXXXXXX		
Tunisia				XXX	
North Carolina				XXXXXXXXXXX	

†Relative solubility index (North Carolina = 100%).

and iron. Therefore, they are not water-soluble phosphates that require biological activity to become available for plants, which is a slow and gradual process. These phosphates liberate soluble phosphorus when they chemically react by contact with weak organic acids generated during the processes of decomposition of organic matter of the soil. Among the important acids are: citric, tartaric, formic, malic, acetic, and other acids generated in the process of humification and mineralization of the organic matter.

McLean and Logan (CORFO, 1986) have established that soils with high phosphorus retention have a lower phosphorus content in plant leaves when the water solubility of the monocalcium phosphates increase. They observed that, in weakly acid soils, the acidulation of 20% phosphate rock improved plant nutrition in several crops with respect to 100% acidulated phosphate rock. This reiterates the need to use a mixture of monocalcium phosphates with phosphate rock in soils rich in aluminum and with low pH.

The efficiency of the tricalcium phosphates and the rate of availability of soluble phosphates is strictly linked to biological activity generated by the crop residue. For this reason, phosphate rock can be used for pasture and crops, especially on trumaos soils, with no-till when the surface residue is permanently on the soil.

Sanchez and Uehara point out that in soils of high phosphorus fixation, the direct applications of phosphate rock in acid soils are more effective and economical than applications of superphosphates. They also mention that the phosphate rock is reactive in acid soils and the cost is a third to a fifth less per unit of phosphorus compared with superphosphate. To satisfy the phosphorus demand of the crops, they point out it may be necessary to apply phosphate rock combined with superphosphate (CORFO, 1986).

The African phosphate processed in Uruguay, named hyperphosphate contains 28% P_2O_5, 46% CaO, 1.4% N, and 1.4% S. This is converted into a granular fertilizer and has a good potential for acid soils planted by no-till or for pastures. This fertilizer contains 9% phosphorus, soluble in ammonium citrate, or 12% soluble in citric acid, or 22% if the solvent is formic acid. The solubility of this phosphorus treated with 5% ammonium sulfate is directly related to the pH of the soil and the quantity of the crop residue on the soil surface. If enough organic matter exists in the soil, there are many possibilities to solubilize the hyperphosphate. Walter Hill (1976) stated that the hyperphosphate is not soluble in water like the phosphates in general, but would be soluble when it comes in contact with a required concentration of organic acids. We know that citric acid starts to solubilize up to 12% of the phosphate rock each time the citric acid concentration reaches 2%. The entire fraction of phosphate rock resists going into solution until the citric acid concentration reaches 100%. This is many times greater than what the biological activity can generate from organic acids.

The determination of the solubility of phosphorus made in the laboratory indicates a concentration specific to the solvent used and the time of agitation in the laboratory container. The conditions of solubility in the laboratory are distinct from conditions in the field because there are different adsorptions due to natural and agronomic management that the farmer can produce in the soil. These indices of reactivity and availability of phosphate are dependent on the fact that the phosphate material is less than 100 mesh in particle size. This last statement is important because the smaller the particle, the better the specific surface of contact for phosphorus–water–soil–plant complex, which favors the soil reaction and phosphate availability.

It is important to point out that total phosphorus in the soil can be made soluble, depending on the levels of organic matter and biological activity of the soil, plus the other factors already mentioned. This explains why the phosphate rock gives good results in the fertilization of no-till pastures and soils with pH less than 6.0. The availability of phosphorus will be slow and gradual. The pH should be checked annually and phosphate rock applied in order to decrease the use of monocalcium phosphate. This also will supply adequate phosphorus until phosphate rock reaches a level of equilibrium with the requirements of the plants.

Commercial phosphate rock is available from Virginia, North Carolina, and Sechura, Peru. The Peruvian product can contribute 30.5% phosphorus, 0.1% potassium, 47% calcium, 4.3% sulfur, and 0.6% magnesium. The index of reactivity is 78%, which falls in the group of phosphates suitable for acid soils. These fertilizers are sold in very fine granular form that permits surface broadcast application on pastures and soils that are not plowed.

The phosphate rock from Bahia Inglesa, near the port of Caldera in Region III, is sold in the south-central and southern zones. The level of phosphorus (18% P_2O_5) is inferior to other natural phosphorus available in the market; however, it has been well accepted and works well for many crops.

The high content of CaO in most rock phosphates improves the pH in the root zone. If acidifying granulated fertilizers are applied, added rock phosphate is a direct benefit for plant nutrition.

2.14. CHLORIDES AND NO-TILL

The high levels of fresh organic material that crops generate under no-till demand adequate management. As mentioned, crop residue must be uniformly chopped and spread on the soil surface to rectify the carbon/nitrogen ratio without impeding the nitrogen availability to the established plants. In many circumstances, especially after the third year of no-till management, rapid decomposition of the crop residue is desired so that valuable nutrients in the residue are readily available for plant growth. Chlorides can help the decomposition process while improving some nutritional aspects and disease tolerance in important crops, such as wheat and barley.

2.14.1. Chlorides and Organic Matter Decomposition

Our traditional agriculture does not mention chlorides (Cl^-) as a nutrient important in soil fertilization, but as a phytotoxic element in some crops. Nevertheless, the application of potassium chloride (muriate of potash) is particularly beneficial in the mineralization of organic matter, a process that is accelerated with the benefit of improving nutrient availability (Goos, 1987).

In the processes of mineralization of organic matter, one of the most important phenomena that occur is ammonification. This is where organic nitrogen is converted to ammonium (NH_4^+)

that is assimilated by plants. Goos points out that chlorides in acid soils can inhibit the processes of ammonium nitrification.

2.14.2. Chlorides and Plant Nutrition

Although chlorides are considered important in the nutritional balance of plants, rarely is this element deficient in the soil. A level close to 0.3 parts per million in plant tissue is considered optimum, which could be estimated low if compared with the need for fluoride, copper, zinc, and manganese (Soil Improvement Committee, 1973). Chloride directly benefits plant nutrition and serves as a catalyst in other processes, such as osmotic movement, stomate action of the leaves, photosynthesis, and enzymatic activity (Goos, 1987).

2.14.3. Chlorides and the Control of Disease in Cereals

Chlorides also benefit plants by controlling root and leaf diseases of such crops as wheat and barley.

Studies in Oregon and North Dakota show that wheat yields increase with the application of potassium chloride, mainly because of a better control of root diseases, such as foot rot caused by the fungus, *Gaeumannomyces graminis* var. *tritici* and *Fusarium* sp. and a positive foliar control of *Septoria*.

Paul E. Fixen at the Potash and Phosphate Institute explains that chlorides in the soil inhibit the transformation of ammonium to nitrate without lowering the pH in the root zone. This helps control diseases. The pathogenic organisms are decreased by the action of microorganisms developed voluntarily upon achieving a pH closer to neutral (Fixen, 1987).

2.14.4. How Much Chloride to Apply

Oregon State University indicates that applying 40 kilograms per hectare (36 pounds per acre) of potasium chloride to winter wheat with the planter, and 80 kilograms (72 pounds) in spring can control root diseases. The response depends on soil type, climate, and distance from the ocean. Fields close to the coast can receive chlorine from rainwater; therefore, the condition of this element in the soil should be known before it is applied. A high seasonal precipitation can leach it. The precipitation of chlorine prevailing from the ocean can reach up to 1 kilogram per hectare (0.9 pounds per acre) each year in areas remote from the ocean; however, this quantity can be elevated several times in areas closer to an ocean.

BIBLIOGRAPHY

Asociación Amigos Del Suelo. 1987. Resumen de 30 años de trabajo. Memoria 1957–1987. Buenos Aires, Argentina. 48 pp.

Agromax S.A. Revisión bibliográfica sobre fosfatos naturales. Departamento Técnico Agromax S.A. 24 pp.

Allison, F. 1976. Crop residues management systems. ASA Special Publication 31:114–121.

Bear, F.E. 1963. *Química del suelo*. Traducción de José de la Rubia Pacheco. Ediciones Interciencia. Madrid, España. 435 pp.

Bernardi, C., E. San Martin, E. Meléndez, and S. Aomine. 1973. Nitrogen and organic matter of volcanic ash soils of Santa Bárbara Association in Chile. *Soil Science and Plant Nutrition* 19(3):139–146.

Besoain, E., y G. Sepúlveda. 1985. Minerales secundarios en suelos volcánicos de Chile. Ira. Edición. Juan Tosso, Editor. INIA. pp. 203–204

Blevins, R.L., L.W. Murdock, and G.W. Thomas. 1978. Effect of lime application on no-tillage and conventionally tilled corn. *Agron. J.* 70:322–326.

Bornemiza, E. 1965. *Conceptos modernos de acidez del suelo*. Turrialba (Costa Rica) 15(1):20–24.

Campos, V. 1983. Control de la germinación de malezas en sistemas cero labranza por acción de compuestos fenólicos simples. Universidad de Concepción, Fac. de Ciencias Bílógicas y de Recursos Naturales, Depto. Botánica. 17 pp. (Tesis de Grado).

Cargill Chile Ltda. 1988. *Manual de fertilidad de los suelos*. Potash & Phosphate Institute. la impresión en español. pp. 11-12.

CORFO (Chile). 1986. Evaluación agronómica de la roca fosfórica de Bahia Inglesa. Gerencia de Desarrollo, Comité de Sales Mixtas. 6 pp.

Crovetto, C. 1983. Importancia de los residuos de cosecha en la producción agricola. *Próxima Década* (Chile) 2(14):13–15.

Crovetto, C. 1986. Cero labranza, extraordinaria alternativa para el cultivo de cereales en suelos erosionados. En Diálogo XV, Sistemas de Labranza y Conservación de Suelos. IICA. Montevideo, Uruguay. pp. 135–144.

Crovetto, C. 1987. La cero labranza en siembras de trigo y su influencia en el medio edáfico en suelos erosionados de la Cordillera de la Costa de Chile Central. *En* Seminario de Producción de Trigo. Universidad Católica de Chile, Fac. Agronomía, Depto. Ciencias Vegetales. Temuco (Chile). pp. 3.0–3.21.

Crovetto, C. 1988. ¿Qué pasa cuando se queman los rastrojos? *Chile agricola* 13(138):288–290.

Demolon, A. 1965. *Principios de agronomia*. Tomo I. Dinámica del suelo. Traducción de la 5ta. edición francesa por José Pérez Malla. Ediciones Omega S. A. Barcelona, España. 527 pp.

Demolon, A. 1967. *Principios de agronomia*. Tomo II. Crecimiento de los vegetales cultivados. Traducción de la 5ta. edición francesa por José Pérez Malla. Edición Revolucionaria. La Habana, Cuba. 587 pp.

Donahue, R.L., R.W. Miller, y J.C. Shickluna, 1988. *Introducción a los suelos y al crecimiento de las plantas*. Traducción de la 4ta. edición en inglés por Jorge Peña C. Impresiones Editoriales S.A. Mexico, D.F. pp. 244–261.

Elliott, L.F., T.M. McCalla, and A. Waiss, Jr. 1978. Phytotoxicity associated with residue management. *In* Crop Residue Management System. ASA Special Publication 31:131–146.

Emesley, J. 1982. A fixation with phosphate. *New Scientist* 30:915–917.

Espinoza, W. 1982. Análisis de las investigaciones sobre fósforo en suelos volcánicos chilenos (1953-1973). Universidad de Concepción, Escuela de Agronomía, Depto. Suelos. Chillán, Chile. *Boletín Tecnico de Suelos* No. 56. 39 pp.

Fixen, P. 1987. Chloride fertilization. *Crop and Soil Magazine* 39(6):14–16.

Garavito, F. 1979. *Propiedades quimicas de los suelos*. 2da. edición, IGAC, Bogotá, Colombia. pp. 91–115.

Gaucher, G. 1971. *Tratado de pedología agrícola. El suelo y sus caracteristicas agronómicas*. Traducción del francés por José Pérez Malla. Ed. Omega S.A. Barcelona, España. pp. 319–322.

Gavilán, J.F. 1981. Efecto de los herbicidas paraquat, atrazina, y Roundup sobre algunas propiedades biológicas del suelo. Universidad de Concepción, Facultad de Ciencias Biológicas y de Recursos Naturales, Depto. Microbiología. 64 pp. (Tesis de Grado).

Gilbey, D.J. 1982. Weed research pays off. *Journal of Agriculture* (Australia) 23(3):81.

Goos, R. 1987. Chloride fertilization. *Crop and Soil Magazine* 39(6):12–13.

Hill, W. 1976. Hiperfosfato. Fosfato natural de aplicación directa. Depto. Técnico Agromac S.A. 6 pp.

Kimber, R.W.L. 1967. Phytotoxicity from plant residues. I. The influence on rotted wheat straw on seedling growth. Australian J. Agric. Research 18:361–374.

Lavandos de Uteda, B. 1989. Cultivo de *Azotobacter sp.* en suelos de Visingso y Torfolk (Suecia). Asociación de Amigos del Suelo. *Boletín Informativo* (Mayo), Buenos Aires, Argentina. pp. 6–7.

Letelier, E. 1967. *Manual de fertilizantes para Chile*. Banco del Estado de Chile. Editorial del Pacifico S.A. p. 43.

Molina, J.S., y L.S. Spaini. 1949. Coloides producidos en la descomposición aerobia de la celulosa y su influencia sobre la estructura del suelo. *Revista Argentina de Agronomía* 19(1):33–49.

Montenegro, B., E. Besoain, y E. Contreras. 1986. Evolución agronómica de la roca fosfórica de Bahia Inglesa. CORFO, Comité de Sales Mixtas. 20 pp.

Muñoz, A. 1990. Mejoremos la fertilidad de nuestros suelos. *Chile Agricola* 15(164):68–69.

Parr, J.F., and R.I. Papendick. 1976. Factors affecting the decomposition of crop residues by microorganisms. *In* Crop Residue Management Systems. ASA Special Publication No. 31. pp. 101–129.

Pauling, L. 1958. *Química General.* 4ta. Ed. Traducción del inglés por José I. Fernández Alonso. Ed. Aguilar. Madrid, España. 694 pp.

Perez, R. 1980. Alelopatía. Interacciones químicas entre plantas. *Revista Creces* (Chile) 6:16.

Pratt, F. 1961. Effects of pH on the cation exchange capacity of surface soils. Soil Sci. Soc. Amer. Proc. 25(2):96–98.

Primavesi, A. 1984. Manejo ecológico del suelo. La agricultura en regiones tropicales. 5ta Ed. El Ateneo. Buenos Aires, Argentina. pp. 146–182.

Robinson, G.W. 1967. Los Suelos. Traducción de la tercera edición inglesa por José Luis Amorós. 2da edición. Ediciones Omega, S.A. Barcelona, España. pp. 219–226.

Romheld, V. 1986. Variaciones en el pH de la rizósfera de varias especies de plantas cultivadas en función de las aplicaciones de elementos nutritivos. *Revista de la Potasa* No. 12. pp. 1–8.

Russell, E.J., y E.W. Russell. 1967. *Las condiciones del suelo y el desarrollo de las plantas.* Traducción de la 8va. Ed. inglesa por Gaspar González y González. Edición Revolucionaria, La Habana, Cuba. 771 pp.

Selke, W. 1968. *Los abonos.* Traducción de la 4ta. Ed. Alemana por Ortwin Güenther-Leon. Editoria Academia. León, España. 441 pp.

Soil Improvement Committee. 1973. *Western fertilizer handbook.* California Fertilizer Association. 200 pp.

Soza, R. 1980. La cero labranza en el cultivo del *maíz. Tecnología y Agricultura* (Chile) 2(9):17–24.

Thomas, G. 1986. Mineral nutrition and fertilizer placement. *In No-tillage and Surface-tillage Agriculture: The Tillage Revolution.* Milton A. Sprague and Glover B. Triplett, John Wiley and Sons, Inc. pp. 93–116.

Thompson, L.M. 1965. *El suelo y su fertilidad.* Versión española por Ricardo Clará Camprubi. 3ra Ed., Editorial Reverté S.A. Barcelona, España. 409 pp.

Thorup, R. 1984. *Agronomy handbook.* The Fertilizer Division, Chevron Chemical Company, San Francisco. pp. 28–29; 201–204.

Tisdale, S.L., W.L. Nelson, and J.D. Beaton. 1985. *Soil fertility and fertilizer.* Macmillan and Co., Ed., 4th Edition, New York. pp. 662–667.

Urbina, A. 1982. Economia del nitrógeno en suelos de cenizas volcánicas. *En* Primera Reunión de Especialistas en Suelos Volcánicos. Universidad de Chile. Fac. Ciencias Agrarias, Veterinarias y Forestales. Dept. Ingeniería y Suelos. Santiago, Chile. *Publicaciones Misceláneas* No. 14. pp. 55-87.

Vivaldi, A. 1990. Chile, un territorio frágil. Edición Especial Diario *El Sur*, Concepción (Chile), 29 de octubre. pp. 4–5.

Zunino, H. 1982. Materia orgánica en suelos chilenos derivados de cenizas volcánicas. *En* Primera Reunión de Especialistas en Suelos Volcánicos. Universidad de Chile, Fac. de Ciencias Agrarias, Veterinarias y Forestales, Depto. Ingeniería y Suelos. Santiago, Chile. *Publicaciones Misceláneas* No. 14. pp. 41–53.

Zunino, H. 1983. Suelos ecuatoriales. Ecologia microbiana, acumulación de humus y fertilidad en suelos alofánicos. Sociedad Colombiana de la Ciencia del Suelo. XII(1):23–28.

CHAPTER 3

Factors That Limit and Favor No-Till Production

3.1.1. Moisture and Compaction

In Chile, annual precipitation increases from north to south and is concentrated in winter. The south-central part has a 6 to 7 month moist period and a 5 to 6 month dry period with scarce and irregularly distributed precipitation. The growth of crops in summer is during precisely the season of least precipitation, the period in which the greatest evapotranspiration occurs; therefore, these crops need irrigation. Our experience during the first 3 or 4 years of managing no-till indicates to us that the lack of moisture as much as the excess can be harmful for crops compared with soils under tillage. When farmers seed or plant during the dry periods, they must have water available for irrigation. This needs to be planned for beforehand. In southern Chile, it is common to see soils with excess moisture during the winter. This excess moisture results from the topographic position, type of soil, or lack of adequate drainage as a consequence of the high precipitation (see Fig. 9, Chapter 1).

Crops are damaged on soils that flood easily or have a high water table. Native pasture can overcome excesses of moisture because the native plants are accustomed to that environment. Under these conditions the value of the energy and protein that the pasture can provide is low.

A seeded pasture or grain seeding in soils with excess moisture would not achieve the benefits expected because plants adapted to that environment would quickly reappear in place of the desired crop.

Soils that remain moist for long periods are commonly clayey. Being fine-textured, they retain more water than coarser soils. In fine-textured soils, the spaces between particles are occupied by water, which causes a lower oxygen content compared with loam soils. This situation worsens when the space is filled with water. In this case, the soil is saturated. **Wet soils or those with a water table at the surface are not suitable for the establishment of no-till**. In no-till it is always advisable to avoid planting or other agricultural traffic when the soil has excess moisture. Recuperating the original microrelief is difficult without tillage.

In the first few years of no-till, the soil increases in bulk density, in other words, the soil tends to compact. This is because roots and mulch, especially mesofauna, take some time to generate and construct natural conduits and pores in the soil, which improve water infiltration and percolation. This can be an initial disadvantage, especially if the soil does not include crop residue. A soil that is saturated for lack of percolation remains wet for a prolonged time, consequently, the crop is harmed. This situation is worse in fine-textured soils, such as the clays of Chequén.

The loss of the natural oxygen content of the soil not only seriously affects plant nutrition, but also the soil microbiology. Under these circumstances, the aerobic microorganisms of the soil, that is, all the organisms that require oxygen for survival, are reduced in population.

An excess of water replaces the oxygen in the soil, which can result in losses of nitrate nitrogen. The microorganisms convert nitrate (NO_3^-) in the absence of atmospheric oxygen to molecular nitrogen (N_2) contained in the nitrate, a phenomenon technically called denitrification. This loss of nitrogen is irreversible so the use of nitrate fertilizers should be avoided when these conditions are present. This is one reason why rice cultivated under water is fertilized with urea instead of nitrate.

Because of all of the problems stated, winter seeding on bottomland soils that have poor drainage is avoided on Chequén; however, as soon as crops like corn are harvested (April), on this type soil, oats are immediately broadcast seeded in the corn stubble. The corn stubble is chopped immediately after fertilization and broadcasting the oats to cover the seed. The object of this seeding is to maintain the soil as biologically active as possible in the period of excess rainfall and low temperatures. If this is done, it achieves an adequate green vegetative cover, greater consumption of moisture, and better absorption of nitrogen fertilizers before they can be leached by excess moisture. This is important in the development and yield of crops to be seeded in the spring (October).

Because level soils have poor drainage in humid areas, soil improvement to correct the drainage problem before the initiation of no-till was carried out using available technology.

Initially, mechanical cultivation of the soil favors the formation of macropores, which promotes root and vegetative development of the plants, but this effect lasts only during the first few months following tillage. **Compaction, and therefore the reduction in porosity of soils is initiated after seeding.** A tilled soil with little vegetative cover will be compacted by raindrops and by passes of agricultural machinery during the seeding and development of the crops. The only defense against soil compaction is to avoid tillage and maintain residue on the soil surface.

3.1.2. The Plow Pan

The plow pan appears to be a common problem among farmers that till their soils, and it also is an obstacle in the development of no-till seedings. A plow pan can be acute at the outset of no-till. It might be advisable to check with a specialist for the need of chiseling or subsoiling prior to establishing no-till.

Plow pans are a phenomenon typical of soils under intense tillage. This dense layer forms gradually by the accumulation of very fine materials, such as silts and clays that fit tightly together and in time harden. The plow pan impedes the normal flow of water and air. This layer is located at variable depths, depending on the penetration by tillage tools. The plow share helps create the plow pan because, by mechanical design, it puts pressure on the bottom of the furrow, generating a layer or horizon of greater density.

According to what I have observed, I think that the formation of the plow pan is initiated the moment at which the soils are plowed or tilled and rainfall occurs. The soil, in the process of drastic alteration of the structure to the depth of impairment, generates two profiles. Each profile has different physical conditions. The disturbed part of the soil is left altered, losing its original structure and at the same time acquiring different physical properties. The new properties include greater initial capacity to absorb water and greater ease of movement of water in the profile. Below the tilled portion, the soil remains without alteration of its natural physical condition. The water absorption capability is much less than that of the tilled profile above.

The tilled profile can store more water as long as the soil is not compacted by the action of raindrops. In practice, the sealing of this profile by the disturbance caused by the raindrops on the bare and tilled soil occurs with the first intense rainfall. Before the soil becomes sealed, it probably enters into a period of saturation with water. Under these conditions, the movement of water in the altered soil layer is greater than in the underlying unaltered soil, producing a marked difference in the permeability of the strata. Because the water initially moves more easily in the surface layer, it is capable of carrying fine particles in suspension. The deposition of particles in suspension occurs when the water and transported solids arrive at the interface with the compact layer. Because of the low permeability of the unaltered soil, the colloidal load is deposited on this layer. This phenomenon is technically called illuviation, and it can be defined as the transport of very small particles of soil by gravitational water. Illuviation occurs naturally in soils under conditions where periods of saturation alternate with drainage of water. These conditions are very important in the pedogenetic processes because they allow the intrusion of active materials to the lower horizons in a state of change.

The plow pan not only affects the normal movement of water in the profile, but also the depth of plant root development, which tend to grow superficially. This situation limits even more the availability of water and nutrients in critical periods. This affects the soil each time that it is tilled to the same depth and with similar tillage implements.

This may be one reason the chisel plow has had acceptance, because it can work to a greater depth than that of conventional plows. This allows for cutting through the layer compacted by the accumulation of the colloidal load; however, the repeated passing of this chisel plow at the same depth also can produce a plow pan.

Because of the special physical characteristics of the trumao soils, this phenomenon is more intense in these soils. A plow pan forms in these soils mainly because of the lack of cohesion between its particles and because the soil preparation operations are normally done with dry soil, which leaves the soil very sensitive to later illuviation.

In clayey soil, the illuviation is less intense and less frequent than in the trumao soils. Tillage implements can only work with appropriate moisture. Also, the physical characteristics of clayey soils are not so favorable for a plow pan to form, at least in a relatively short amount of time.

With what has been stated, one can conclude that the plow and other tillage tools are direct causes of plow pans, and are directly responsible for the drastic change in the structure between the tilled soil and the underlying untilled soil. Not plowing and leaving the residue on the soil can be the best alternative to overcome this problem. When no-till is initiated in a soil that has a plow pan, I suggest that a chisel plow be used just once before seeding to loosen the compacted zone.

Compaction is foreign to the natural soil process. People are responsible for the poor management of the soil by not considering the residue and by an intense and inappropriate agricultural mechanization.

3.1.3. Soil Temperature

The presence of mulch over the soil can modify substantially the temperature of its surface. The mulch helps to (a) absorb solar energy and (b) insulate the soil against high and low temperatures. This phenomenon is due to two basic aspects:

1. The color of the mulch can influence the absorption of solar energy. Wheat residue, even when it remains yellow, irradiates part of the solar energy it receives and radiates it toward the atmosphere. This is beneficial because the soil stores less daily energy during the summer compared with a bare soil. In the cold months, the mulch changes to a darker color, which indicates that it is capturing and retaining heat useful for soil processes.
2. The mulch insulates the soil, obtaining lower temperatures in summer and higher temperatures during the winter than those on an uncovered soil (Marelli et al., 1981). Mulches insulate the soil better because different layers of air remain between the accumulated residue (Thomas, 1986).

Farmers that practice no-till are concerned about the lower daily temperature registered in spring compared with that of a bare soil. The explanation of this phenomenon has already been mentioned in previous paragraphs. At the end of winter, the soil is colder because the mulch that covers it impedes a rapid absorption of heat. The opposite occurs in bare or cultivated soils. If the soil has a greater quantity of moisture, the warming of the soil is more gradual, thus planting can be delayed until better temperatures are reached for germination and development of the seedlings. On Chequén, we plant corn 1 week later than in a traditional system. Earlier seedings germinate 2 or 3 days later, and the plants are pale green and less vigorous in their first 15 days.

According to the monthly agrometeorological bulletin of the University of Concepción, Chillán campus, the temperatures in the cold months of winter (1988) are only lower by less than 0.1°C in soils covered with grass compared with bare or cultivated soils. In hot months, the temperature is 3 to 6°C lower in a soil with grass compared with a soil without grass (Fig. 16 and 17). This clearly indicates that maintaining the soil with a mulch cover contributes to achieving a more stable temperature in the soil.

In fall seedings, we have observed that crop residue does not negatively affect soil temperature, so the crops are established in better conditions. This is the reverse of what happens at the end of winter. In fall or the beginning of winter the soil has a higher temperature that is maintained for a longer period by the insulating effect of the mulch.

In the summer, the temperature of the air and soil reach extremes that can harm the normal development of plants. In the soil, the maximum temperature should not exceed 30°C (86°F). The mulch on the soil fulfills an important insulating role to maintain the temperature as well as prevent losses of moisture by evaporation. This translates into achieving a more beneficial medium for microbiological activity and for the established crop.

One can conclude from what we have observed at Chequén that organic residue on the soil generates conditions that favor a more uniform temperature in the profile.

3.2. Soil Acidity

One factor that can most affect the establishment of no-till is soil acidity. The decomposition of crop stubble is done by microorganisms that require an appropriate pH for their work. Soil acidity, together with the lack of organic matter, can be decisively harmful factors to yields. They also are the most important parameters of the soil.

A notable reduction in soil pH has been observed in south-central Chile in the last 20 years. This could be attributed to the following examples of poor soil management:

- Liming is no longer done.
- The soil is tilled resulting in loss of organic matter.
- Massive amounts of ammonia and/or ammonium fertilizer is used.

Before the 1970s, calcium carbonate was commonly used on the soils of Regions IX and X. Liming is no longer a common practice, which has resulted in a gradual loss of the calcium and hydroxyl ions, an increase in acidity, and, consequently, a decrease in soil pH.

Crops extract calcium, magnesium, potassium, and other nutrients from the soil. These nutrients are in the seed and in the residue. If the residue is burned, these bases can be almost completely lost as the rising air currents generated by high temperatures transport ashes rich in nutrients to other places. Thus, winds that affect the soil surface after a burn are an important factor in nutrient losses.

Low levels of calcium and other soil bases, together with the reduction of pH, can affect plant nutrition. The importance of soil microbiology that is managed within a relatively neutral pH should be noted. Bacteria are very sensitive to low pH, which diminishes the potential in the degradation of the stubble. Other microorganisms, such as actinomycetes and fungi, can live in more acid soils, but their activity also can be diminished.

The timely and natural biodegradation of residue on the soil is the basis of no-till. It is vitally important to maintain appropriate levels of nitrogen, magnesium, and especially calcium in the soil, all within a normal pH level.

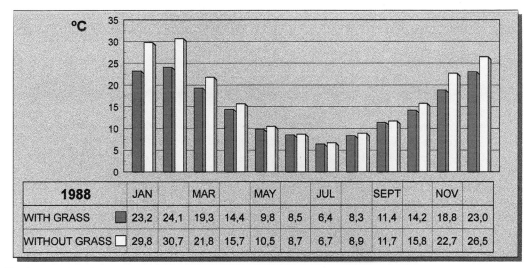

1988		JAN		MAR		MAY		JUL		SEPT		NOV	
WITH GRASS	■	23,2	24,1	19,3	14,4	9,8	8,5	6,4	8,3	11,4	14,2	18,8	23,0
WITHOUT GRASS	□	29,8	30,7	21,8	15,7	10,5	8,7	6,7	8,9	11,7	15,8	22,7	26,5

Fig. 16. Behavior of temperature at the soil surface.

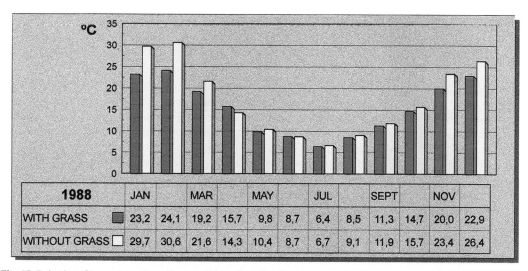

1988		JAN		MAR		MAY		JUL		SEPT		NOV	
WITH GRASS	■	23,2	24,1	19,2	15,7	9,8	8,7	6,4	8,5	11,3	14,7	20,0	22,9
WITHOUT GRASS	□	29,7	30,6	21,6	14,3	10,4	8,7	6,7	9,1	11,9	15,7	23,4	26,4

Fig. 17. Behavior of temperature 2 centimeters below the soil surface.

3.2.1. Tillage, Organic Matter, and pH

As previously mentioned, the tools that cultivate the soil today are rapid and efficient in their work; however, they unleash the erosive processes and permit loss of nutrients by surface transport.

Excessive tillage and the consequent reduction of organic matter levels can mean a loss of the OH^- ion and an increase in soil acidity. Soils that have lost organic matter also have compromised their structure. This leaves them subject to leaching of useful cations (Ca^{+2}, Mg^{+2}, Na^+, K^+, and NH_4^+) by intense rains or prolonged irrigation, which also can lead to an increase in the soil acidity.

A permanently managed soil without tillage tends toward its original pH as long as the use of ammoniacal nitrogen is countered by the application of calcium carbonate or is replaced by the use of nitrate nitrogen.

In clayey or silty soils, a reduction in the levels of organic matter also can be synonymous with a reduction in pH. When the existing organic matter on the soil surface is buried, a more rapid process of decomposition is initiated than if it were to remain on the surface. In these processes, initially, the pH can change and tend to acidify because of the active generation of organic acids. When the stubble remains over the soil, the decomposition is slower, therefore, the possibility of altering the pH is less.

The soils with higher organic matter content have a greater buffering capacity; they can better conserve their original pH. This is very helpful for farmers, so care must be taken to maintain the organic matter level.

Farmers in Chile tend to use ammoniacal fertilizers, especially urea. As mentioned in previous section, more urea is used each day because of the price difference per unit of nitrogen compared with nitrate nitrogen. This can be important if the acidifying effect of urea and ammoniacal nitrogen in general is considered. **The decision to purchase these fertilizers should be looked at for its agronomic performance as well as the economic aspect.**

In conclusion, soils that have low organic matter content, and the greater possibilities of acidification that goes with this, can make the establishment of plants in no-till more difficult. The main reasons for this are less availability of nutrients, reduction in microbiological activity, and the presence of toxic elements. All these situations are aggravated by tillage.

3.3. Nitrogen

Nitrogen is an indispensable chemical element in managing no-till. I believe that many failures in the system have been caused by not considering the carbon/nitrogen ratio. To correct this imbalance, more nitrogen must be applied than the crop needs. The additional nitrogen application should not be considered an added expense, but a necessary investment because in addition to obtaining better yields, a soil building stage will be initiated. Undoubtedly, some of this phenomenon represents improved distribution of other basic nutrients, such as phosphorus, potassium, and calcium. Enhanced residue decomposition is caused by increased biological activity stimulated by higher initial application of nitrogen.

Plantings into small grain residue tend to need more nitrogen during the first 3 to 5 years after initiating no-till. The nitrogen requirements will decrease as the system reaches an equilibrium with the nitrogen released by mineralization of organic material and formation of humus. The rate of residue decomposition can vary by the type of soil, the type and amount of residue, temperature, ambient humidity, and manner of production.

Rates of nitrogen applied at planting should be increased for about the first 5 years. The reason for this is that during the crops initial development period, surface applied nitrogen can be trapped by the residue and can cause deficiencies in the seedling of the planted crop. The first residue that accumulates on the soil surface when a no-till system is started releases nutrients slowly. This is very important since the release of nutrients from organic matter, especially the release of nitrogen, occurs gradually, benefiting the crop and avoiding unnecessary nutrient losses.

Burning of residue can undoubtedly prevent extensive nitrogen immobilization; however, one cannot ignore the fundamental importance of organic material in the intermediate and long-term soil conservation and improvement of physical, chemical, and biological soil properties. It is well worth the effort to spend a little more money to initiate no-till and later reap its notable benefits.

3.4. Diseases and Pests

3.4.1. Diseases

No-till is strongly promoted by farmers convinced of the system because it allows them to ensure a more fertile, erosion-free soil. At the same time, however, the farmers proceed with caution. They typically decide to abandon the plow and harrow after having seen their neighbors successes or failures. They applaud the successes, and probably use the failures to find solutions for their mistakes.

Undoubtedly, some factors that generate resistance to using no-till are diseases and pests. Not burning the residue and leaving it on the soil surface causes farmers to think that the established crop can be affected by pathogens resulting from previous residue; however, these problems are easily avoided with adequate treatment and appropriate crop rotations.

The less humans interfere with the soil, the greater the possibilities of improving its ecological equilibrium. It is possible to overcome diseases and pests. Mother nature creates residue and leaves it on the soil, where it forms the fundamental resource for our work. Of all the agricultural systems man has created, **no-till emerges as the best restorer of the biological processes that form the basis of life. Examples are seen in forests and other natural, unmanaged fields.**

The least biologically disturbing form of planting that obtains the objectives of soil conservation and crop production is one in which no more than a drill or planter is used. The operation places the seed and fertilizer in the residue of the previous crop with minimal trips across the surface.

At Chequén, we have had success with disease control especially in wheat. Even in monoculture we have achieved excellent control of the take-all *Gaeumannomyces graminis*, a disease endemic to most of the soils under limited tillage in the south-central part of Chile (see Chapter 4).

A corn–wheat rotation has demonstrated benefits in controlling diseases without becoming a major problem in the management of these cereals, regardless of the planting method. Folicur Bayer is applied in the spring to control *Puccinia* sp. This is applied in addition to the normal seed disinfectant. Foliar diseases, such as septoria, have been controlled in their early stages simply with nitrogen fertilization. This indicates that nitrogen application should be made early and in adequate amounts. Incorporation of *Lupinus angustifolius* L. in the corn–wheat rotation lengthens the rotation and creates a natural medium that is richer in nitrogen. This helps control common diseases and improves productivity of the system.

3.4 2. Pests

3.4.2.1. Slugs

Legumes are more sensitive to attack by larvae and slugs than are cereal crops. Slugs have damaged the seeds and stems of lentils, garbanzos, vetches, soybeans, lupins, and alfalfa planted at Chequén. We established a careful study of the different pests that attack crops. We used specific pesticides applied to the seeds, sprayed directly on the foliage, or in the form of bait, which is quickly consumed; however, these controls were not entirely effective, especially with slugs, and they present problems for establishing legumes in general and corn in particular.

Science has developed sufficient technology to control most pests that affect crops. Obviously, this development has primarily surfaced in areas that demanded control. There is probably little research and consequently scant understanding of control of pests that do not surface as real problems in traditional agriculture. An example of this is the lack of research on control of slugs (*Agrolimax reticulatus*). For soils under tillage, slug existence does not threaten crops because the effect of tillage or use of fire on residue considerably reduces their natural habitat. Every time the soil is disturbed, some slugs remain on soil surface facilitating natural control by predatory birds.

Slugs, a mollusk, have seriously damaged shoots and leaves at Chequén in the first stage of plant development. They can be controlled by applying a baited mollusk poison, such as methiocarb or metaldehyde, at rates recommended on the label. The commercial product, Mesurol by Bayer, controls slugs only after two or three applications.

Slug eggs in no-till soil (1989).

Although the habit of adult slugs is nocturnal, they may also damage crops on humid, overcast days (1989).

Undoubtedly, the residue and the humidity retained because of the residue stimulate prolif-eration of slugs. Controls should be applied with the first autumn rains and in the spring before planting. This semiannual application only helps decrease the intensity of the pest. If damage per-sists after crop emergence, it will be necessary to determine if the pest is localized or generalized, and then proceed with its control. The control measures should be repeated until the damage is prevented.

Mollusks get their nourishment by scraping the tender green tissues with a mouth part, called a radula. This facilitates identification of the pest. The affected tissues show strip removal with damage in the center of the leaves. This weakens the tender plants, reducing their vigor or destroying them. Slugs can be identified by the bright, filamentous waste trail. The secretion is left on the surface of the green tissues as well as on the soil.

In different legume plantings, we noticed that the intensity of slugs varied widely from one field to another. The damage ranged from almost imperceptible in some soils to total loss in oth-ers. This shows that keeping track of each field is very important. The type of soil, crops planted, rotations, and management of the residue and the soil itself must be known.

The areas most affected by slugs are those under irrigation, especially those in corn, soy-beans, and spring lupins. In corn plantings, the damage appears on germinating seeds and seedlings with the first leaves scraped at the center. Plants broken as a result of weakening the stem are also observed. Even though the damage can be quite significant in spring during the first 40 days of development, corn has an exceptional capacity to overcome the damage if it is ade-quately managed for rapid growth.

Most pests that affect no-till crops are hidden in the residue, so that they are protected from their natural predators and are not easily seen. These pests can only be identified at night or at dawn when they are fully active and are more easily identified and counted. Their damage also can be assessed at this time.

3.4.2.2. The Bean Fly (Seed Maggot)

As mentioned in Section 1.4, the bean fly (*Hilemia platura*) can be a problem for establishing legumes in no-till. The adult bean fly, like the common housefly, deposits its eggs on residue left after harvest. Yellowish, white larvae hatch and can penetrate the germinating seed. They attack mainly in the spring and can destroy an entire planting. The damage is recognized by chewed seeds and cut, new stems. These pests can be controlled with specific pesticides applied to the foliage and systemic nematicides applied to the root (Gerding and Quiroz, 1987). Good control of the larva has been obtained by mixing the seeds with insecticides, such as Chlorpirifos (50%) or Carbofuran (44%), in a revolving drum at rates recommended by the manufacturer. At Chequén, the bean fly has caused problems in corn, such legumes as soybeans, and lupins. The controls used have been promising, especially in lupins.

3.4.2.3. Crane Fly

In soils recently converted to no-till, we have detected infestations of *Tipula apterogyne Philippi* on small clusters of plants. Although the damage in the affected areas was total, the areas were small. In accordance with the observations of Dr. Roberto González (1989), this Diptera, belonging to the suborder Nematocera, has an aquatic or subterranean habit and phytophagus characteristics. These observations support the development of this larva in soils that are humid and have a low level of organic matter at the start of a no-till system.

Cocoon and caterpillar of the stem cutworm. This is not a major pest in no-till (1991).

Crane fly larvae are active in seedbeds and fields during autumn and winter. In no-till they cause only spot infestations that are easily controlled.

Adult crane flies, also called daddy long legs, are observed in wheat in November (1990).

White grubs do not cause damage in no-till (1991).

In soils managed with abundant residue, no damage has been observed in the crops indicating that this larva, when it does not encounter sufficient residue, can seriously damage established plants. This problem can be accentuated when weed control is effective. An appropriately timed treatment with a contact insecticide applied to the foliage, such as 44% carbofuran at appropriate rates, gives good results to the infested areas.

3.4.2.4. White Grubs

We have noted the presence of white grubs since the beginning of no-till plantings at Chequén. After 12 years of management without the use of soil disturbing implements, we have seen neither a decrease nor an increase in the population of these larvae. They belong to various species of bloomers, Coleoptera of the Scarabeidae family. We have learned from Embrapa Agricultural Experimental Station in Passo Fundo, Brazil, that white grubs, *Bothynus* sp. attack root plants only when they do not find enough crop residue on the soil surface. This research is coincident with our experience showing no damage from white grubs when straw is kept on the soil surface. This coleoptera prefers straw rather than fresh roots, but if it does not have straw, it will eat green plants. This outstanding example clearly shows benefits in pest control through no-till.

In summary, one could say that no-till demands a greater control of pests, especially in legume plants. Application of specific pesticides may be necessary initially.

3.5. MONOCULTURE

In many cases, farmers practice monoculture of a particular crop with relative success. They often do this because they lack the knowledge needed to introduce other crops. Planting wheat on wheat stubble with no-till can have clear advantages over tilled soil, but, as I have noted in Section 3.4.1. plantings of wheat on wheat stubble wears out the soil more quickly on plowed soils than on no-till soils. Even though monoculture can produce some short-term economic benefits, one should consider the plant health aspects of fertility and ultimately, of production in the medium and long-term. In this respect, it is important to note that fertilization of the same crop and control of pests and diseases repeatedly use the same agrochemicals. This can cause a serious deterioration in the local ecosystem because these chemicals can systematically develop forms toxic to other biological systems. For example, continued use of atrazine in monoculture of corn can affect other crops and contaminate water bodies through overland flow or percolation. This will in turn contaminate drinking or irrigation water. The same situation can occur with the excessive nitrogen fertilizers used in a corn monoculture.

Literature indicates that continuously applying the same pesticides induces resistance to the pathogens and at the same time destroys their natural enemies, which in the extreme case can economically destroy the crop. The same situation can occur with weed control, which will encounter serious difficulty when weeds that previously were secondary to control under monoculture will be more difficult to control. The fact that many plants do not do well under monoculture indicates that nature imposes its natural laws, showing us other alternatives.

On the other hand, crop rotations ensure production with lower costs and better results. These benefits are probably intensified under no-till, and production goals are reached in a more efficient time frame under the context of a conservation system. Figure 18 shows the chronological and management scheme for a corn–wheat rotation under no-till carried out at Chequén. Lupins and wheat are in the rotation every 3 years.

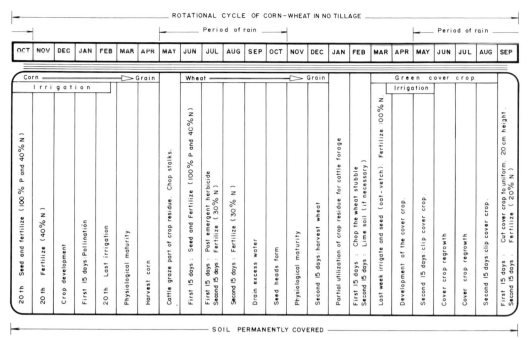

Fig. 18 A chronological scheme and management of the wheat–corn rotation with no-tillage at Chequén.

3.6. BETTER SKILLS IN MANAGEMENT

The adoption of new technology in agriculture, to put it simply, requires a better understanding of the parameters of crop production and, particularly, of the fundamental resources. All this signifies a greater dedication by the farmer to strive to obtain the goals and success he desires. Generally, those that innovate will earn better profits and gain better knowledge; however, it also indicates that any error in managing the system can create an economic setback that can affect the farmer to a small or large extent.

When the decision is made toward conservation of the soil, then the search begins for better ways to treat the soil. I believe that in addition to meaningful things like new technology, also goes anguish, optimism, worry, and happiness. **To initiate the no-till system requires a deep conviction and belief that I am going to make it good, even better than good. Positive thinking is a necessity!**

Many times after we become aware of the effects of soil destruction, other forces appear that bring about changes. We realize that we cannot continue farming the same way our parents and grandparents farmed. If we are firmly convinced to innovate our farms using the natural basis and logic from our studies and observations, we can change the old ways to better methods. We will then promote better treatment of the soil and reap the continuing benefits.

To adopt the no-till system, we must be aware of the important role humans play in the soil resource. If we really instigate this belief, we are going to have success in this new philosophical conception of soil management; however, in the startup of the system, a basic understanding of agronomy, as well as some experience, is needed for an easier transformation. Errors can happen, and some times we will question the planning that went into the method of operation. Yet this does not have to be an obstacle.

Personal observations in the field are essential. The appropriate lessons must be learned in a short time and they must be learned by the people who are directly involved in the production process. It is important to use these observations to verify the work of the field operators in planting, application of pesticides, fertilization, and harvesting.

The no-till system demands a greater skill and precision in all aspects of production because it consists of a real revolution in the management of soil. An open mind is required to understand and then modify whatever circumstances or shortcomings occur. These trials and errors are not a weakness in the system, but more than anything, a contribution to the knowledge and understanding in our search for better solutions. This will lead to future successes.

I believe that my work, like all the agriculturists in the world, is the most important work in the development of mankind.

BIBLIOGRAPHY

Gerding, M., y C. Quiroz. 1987. Control de la mosca del frejol. IPA, INIA, Quilamapu (Chile) 32:16–18.

González, R.H. 1989. Insectos y ácaros de importancia cuarentenaria en Chile. BASF, Ed. Ograma S.A. Santiago, Chile. pp. 221–222.

Marelli, H., B.M. DeMir, y A. Lattanzi. 1981. La temperatura del suelo y su relación con los sistemas da labranza. INTA, Estación Experimental Regional Agropecuaria de Marcos Juárez, Argentina. Informe Especial No. 14, Serie Suelos y Agroclimatologia. 8 pp.

Thomas, G.W. 1986. Mineral nutrition and fertilizer placement. *In No-tillage and Surface-tillage Agriculture: The Tillage Revolution.* Milton A. Sprague and Glover B. Triplett. John Wiley and Sons, Inc. pp. 93–116.

CHAPTER 4

*No-Till and
Its Influence on
Some Natural
Parameters
of the Soil*

A virgin soil profile has a dark color and a high content of organic matter in the upper part.

After 32 years without plowing and 11 years producing no-till corn and wheat in rotation this soil is similar to the virgin soil.

A profile of a soil under traditional tillage shows a lack of an organic surface layer (1991).

I strongly believe that farmers should have as a fundamental goal to work the soil and extract fiber and food without damaging its original functional capacity. To protect the soil against erosion should really be a goal in the life of farmers, especially those that have mainly sloping land. Gravity irrigation is more difficult on slopes, and the soil is exposed to surface runoff by intense rains.

I think that the richness of Chequén is in the first few centimeters of the soil and not what is produced on it. Chequén value is not in the corn or wheat it produces. It is an intangible richness that will continue to increase as long as my descendants continue to follow the path already laid out. As a result, this land will be invulnerable to time and can be enjoyed eternally.

To permit the soil to regenerate naturally, and at the same time achieve the highest production, is pleasingly stimulating and promising. Worn-out soils that have lost their content of organic matter and original structure by erosion can in many cases have good qualities for improvement. Equally, level soils that do not have evident problems, have degradation or wear because of the management system that used traditional tillage elements (plowing, disking, or cultivating) for many years. All this translates into a reduction in the levels of fertility, thereby affecting production. In the same way, these soils can have good qualities for their improvement and will probably be easier to regenerate than soils with slope. Level sandy loam soils merit special mention as they are characterized by having low moisture retention and poor unstable structure. To plow these fragile soils is, without a doubt, a grave error since this further deteriorates their structure and reduces their level of organic matter and fertility.

As I have mentioned before, I believe that the increase in the levels of organic matter has been fundamental in achieving the results obtained at Chequén. The organic matter helps control the erosive processes, even reaching the extreme benefit of initiating the creation of a new soil.

Erosion is controlled and diminished to such a point that one can affirm there is now no erosion on Chequén. The mulch over the soil shields against the impact of the raindrops, which greatly improves water infiltration. If the intensity of the rain does exceed the infiltration capacity, water slowly runs off, hindered by the mulch left on the soil in such a way that a very low quantity of sediment is in suspension.

Increasing the levels of organic matter has been most important in achieving our objectives and goals, as is shown later in a series of soil analyses from Chequén. A mutilated soil has been made productive. The erosive processes in this soil signified a loss of about 1 meter (3.3 feet) of original fertile soil. Its productive capacity was less than 1000 kilograms of wheat or 50 kilograms of meat per hectare per year (893 pounds per acre of wheat or 45 pounds per acre of meat). These overwhelming results indicate that in millions of hectares of soils similar to those of Chequén, productivity would be much greater if the content of organic matter was increased by maintaining crop residue on the surface.

Without a doubt, the only way to achieve this revolutionary change in the natural parameters of the soils of Chequén has been to leave the crop residue on the surface. This required that the traditional tools used to cultivate the soil be eliminated.

In normal pasture management, livestock remove about 70% of what they ingest. The rest is returned to the soil in manure and urine. The manure is concentrated on a small surface and does not contribute a uniform return of the organic matter to the soil. Better distribution of the manure can be obtained by using a tire drag. During winter and summer, livestock tend to look for refuge in the bottoms where shrubs, plantations, or forests exist. This aggravates the distribution problem of the manure and urine, greatly diminishing the proper distribution of fertility. In no-till, efficient management of residue returns about 50% of the total weight of the crop to the soil. More

A drag made of tires is useful in spreading manure and covering open furrows in no-till seedings (1989).

importantly, the high volume that this represents will uniformly cover all the surface. This means that crops return a large portion of the nutrients used in their fertilization to the soil, especially phosphorus, potassium, and calcium. Also, the material protects the soil against erosion and contributes significantly to conservation of soil moisture.

On Chequén, the management of approximately 8 tonnes per hectare (3.5 tons per acre) of stubble from different rotations has meant the increase in thickness of a new organic horizon. This horizon includes everything from black humus to fresh mulch that has not yet been naturally decomposed into the underlying mineral soil.

The analyses of organic matter on Chequén soils show a large change in their levels, compared with the same soils under pasture and traditional cultivation. In the first horizon, 0 to 5 centimeters (0 to 2 inches), soils under no-till for 7 years increased their organic matter content 3.74 times with respect to a soil under cultivation and 17% with respect to seeded dryland meadow. Along with this, these soils were comparatively more efficient in the conservation of the soil and improvement of organic matter. Table 4–1 shows these comparisons.

My intention is to express in simple terms what I have been able to understand as an agricultural producer. I am interested in knowing what is happening in my soils. I consider the management of the stubble, gradual and natural incorporation of the organic material generated by the crops, and their influence in the management of the physical, chemical, and biological characteristics of the soil under no-till, to be critical in achieving sustainability in agriculture.

Table 4–1. Variation in bulk density in relation to organic matter content under three management systems in soils of Chequén.

Management system	Depth	Organic matter	Bulk density
	cm (inches)	%	grams per cubic centimeter
Traditional	0–5 (0–2)	1.42	1.30
Wheat–oats	5–10 (2–4)	1.24	1.38
5 years	10–20 (4–8)	1.00	1.60
Pasture	0–5 (0–2)	4.56	1.05
15 years	5–10 (2–4)	1.92	1.42
	10–20 (4–8)	1.14	1.20
No-till	0–5 (0–2)	5.32	0.95
Wheat–corn	5–10 (2–4)	2.84	1.58
7 years	10–20 (4–8)	2.24	1.60

4.2. IMPROVING THE PHYSICAL CHARACTERISTICS OF THE SOIL

The principal advantage of no-till is the notable improvement in soil physical qualities. In the years that have passed without plowing, the soil is darker, more organic, and softer to the touch than tilled soils. This makes seeding as well as the subsequent development of plants, easier. If the management of the stubble and the rotation of the crops are adequate, these characteristics keep improving each year the soil is no-tilled.

Table 4–1 shows the variations in the bulk density in relation to the organic matter content observed under three management systems on soils of Chequén. The density of a body can be understood as the relationship between its weight and the volume that it occupies. In the soil, this relationship, expressed in grams per cubic centimeter, is the weight expressed in grams of a volume contained in a cubic centimeter. The arrangement and natural aggregation of soil particles, including the spaces or pores with air, are considered because the measurement is made with a dry sample. In soils under a traditional tillage system, the first 5 centimeters (2 inches) of the profile had a bulk density of 1.3 grams per cubic centimeter, compared with 0.95 grams per cubic centimeter under no-till. Therefore, the same volume of soil under no-till weighs less than that under traditional tillage.

Improvement of the bulk density of the soils is directly proportional to its organic matter content. The large increase in biological activity that is gradually achieved with additions of organic matter induces the formation of pores in the soil, increasing the space not occupied by solids.

The water retention capacity also is notably improved with an increase in soil organic matter. This relates directly to improvement of bulk density because the greater volume of space allows more water storage for a longer period than allowed in a soil with lower organic matter content. In practice, one observes better yields in soils under no-till, especially where mulch accumulates on the surface for 2 or 3 years. These high yields are especially evident in years of less rainfall. The values in Table 4–2 relate to available water under three soil management systems of Chequén.

The positive effect of no-till on available soil water is evident by the more than 50% increase compared with a traditional system and more than 35% increase compared with pasture. In practical terms, this means that the same soil managed with traditional systems can store 143 cubic meters per hectare (1.40 acre inches) while in no-till, the available water increased to 232 cubic meters per hectare (2.25 acre inches) in the first 20 centimeters (8 inches) of the soil. Some

Vegetative mulch is a product of the decomposition of residue that is left annually on the soil (1989).

studies indicate that the increase in available soil water under no-till during the summer is greater than that observed in plowed soils (Tisdale et al., 1985; see Fig. 19).

Water that enters the soil from rainfall or irrigation and initially fills all the empty pores not occupied by solids will remain in the soil profile as long as the addition of water at the surface continues. When this addition ceases, a drainage phase begins that corresponds to the downward movement of water in the soil caused by gravity. This phase continues until virtually all the pores large enough to allow free gravitational movement of water have been emptied. Under normal conditions this phase lasts between 24 and 48 hours after the addition of superficial water has stopped.

The water retained in the soil against the action of gravity corresponds to the soil moisture level, called field capacity. It is quantitatively expressed as the percentage of water related to the weight of the soil sample dried in an oven at 105°C (221°F) The water is retained in the pores between the soil particles called capillaries. The finer the pores, the more resistant they are to the mechanisms that tend to eliminate the water that occupies them.

Forces that eliminate the soil water from the pores include absorption by plants for their physiological processes and evaporation that occurs directly from the soil surface. The combined effect of water loss from plants by transpiration and from the soil by evaporation is what is called evapotranspiration. This phenomenon is responsible for the consumption of water held by the soil against gravity.

If there is no addition of water, the point is finally reached at which the absorption of water from the capillaries surrounding the root zone is so slow that the plant wilts. This seriously affects the development of the plant and, in the extreme case, kills the plant. In these conditions, the soil can still have some capillary water, but it can not be absorbed by the plant. This phase is called

Table 4–2. Variation of available moisture in soils of Chequén under three management systems.

Management system	Depth	1/3 atmosphere[†]	15 atmosphere[†]	Available moisture
	cm (inches)		%[‡]	
Traditional	0–5 (0–2)	10.80	5.30	5.50
Wheat–oats	5–10 (2–4)	10.10	5.40	4.70
5 years	10–20 (4–8)	12.60	7.90	4.70
Pasture	0–5 (0–2)	12.80	6.60	6.20
+15 years	5–10 (2–4)	16.70	10.50	6.20
	10–20 (4–8)	17.40	11.80	5.60
No-till	0–5 (0–2)	23.70	15.30	8.40
Wheat–corn	5–10 (2–4)	21.70	13.40	8.30
7 years	10–20 (4–8)	24.40	16.50	7.90

† 1 atmosphere of pressure = 1.033 kilogram per square centimeter.
‡ Dry weight basis.

the permanent wilting point and quantitatively is expressed as percentage of the oven dried weight of the soil.

Available soil moisture is the amount of water that plants can use for their physiological needs and quantitatively corresponds to the moisture between field capacity and the permanent wilting point. In practice, the literature shows that a high correlation exists between field capacity and the moisture content measured in a soil sample subjected to a pressure equal to a third of an atmosphere, as well as between permanent wilting point and its moisture content measured in a sample subjected to a pressure of 15 atmospheres (Gavande, 1973). As is shown in Tables 4–1 and 4–2, the effect of the soil organic matter affects bulk density and available moisture. This is as important for fine textured soils with a dominantly clay fraction as for coarse soils with a dominantly sandy fraction. The effect observed in the first 20 centimeters (8 inches) of the soil profile, in time, will continue to increase in depth, which is precisely where the existence of a soil–air–water equilibrium is desired. Plants will then be in a close to ideal environment, facilitating the complete functioning of the physical, chemical, and biological complex of the soil.

From the preceding comment it also is possible to point out another aspect that is relevant and intimately related to an important physical characteristic, the soil structure and its stability. In

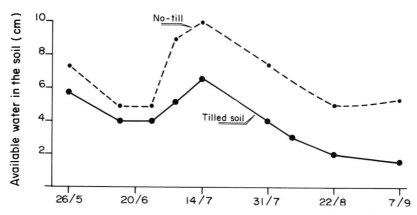

Fig. 19. Availability of moisture in the profile from 0 to 60 centimeters (0 to 24 inches) for corn under no-till and tilled systems (Tisdale et al., 1985).

a soil managed with traditional tillage, the soil structure developed between each cultivation is destroyed when the tillage tools are used in the next cultivation. This increases the potential for leaching, not only of nutrients in the soil, but also the compounds responsible for aggregation of soil particles into larger aggregates that make up the soil structure and characterize the relation between solids and pore spaces. The constant use of tillage implements (plows and disks) leaves the soil without structure. These poorly structured soils are dusty during dry periods, muddy in rainy periods, and subject to erosion and crusting.

Residue incorporated into the soil by the classic tillage tools physically contributes to improve the aeration of the soil. It increases the percentage of pores in the soil and at the same time, impedes the arrangement of the solid particles. The humification processes will only occur insofar as the responsible microorganisms find the nitrogen necessary to begin their binding action. Unfortunately, this process of soil aggregation is made more difficult if it is subjected to another tillage.

These problems in a tillage system do not occur in the no-till system, whose fundamental characteristic is to maintain the structure of the soil that has been able to form through time. Moreover, the stability and improvement of the structure gradually approaches an ideal natural condition. The organic matter that is added by the crop residue in the soil is incorporated as humus in a **gradual and natural way** following the path of porosity generated by activity of roots, microorganisms, and especially of the mesofauna that rapidly colonize the profile of new organic soil. This mechanism, accelerated by the beneficial human action on the environment, is the principal way the physical characteristics of the soil are improved.

4.3. IMPROVEMENT OF THE CHEMICAL CHARACTERISTICS OF THE SOIL

Of all the positive benefits that the no-till system has, the most important for agricultural producers is related to fertility. The reason for this is that fertility directly involves productivity and, consequently, is related to the economy that supports the system.

The analyses of the soils of Chequén, under three forms of management and different horizons, indicate substantial changes in their chemical parameters in seven years. The soils in the traditional system were compared with those in no-till. Nitrogen increased 3.46 times more in a soil that was no-till seeded to a corn–wheat rotation than in a soil using the traditional system. Available phosphorus increased 4.85 times, and potassium 1.55 times. This dramatic change results from the increase in levels of organic matter. This clearly indicates that even with high yields it is possible to increase the fertility level of soils if crop residue is left on the surface.

Nitrogen levels in the soil can be increased relatively easily by applying nitrogen fertilizers in larger amounts than crops require; however, the same does not happen with phosphorus. To raise the level of phosphorus in a soil that is tilled annually is very expensive. Mechanisms that tie up phosphorus are activated by lowering of the pH, primarily because of the loss of calcium by leaching. In no-till, with the use of residue on soil surface, fixation of phosphate is significantly reduced so plants can obtain enough available phosphorus from that normally unavailable to them, even with high crop yields.

For this reason, the quantity of phosphorus applied as a fertilizer on Chequén has been reduced. In recent years, phosphorus has not been applied to seedings of wheat in corn stubble because of the high content in the soil, as shown in Table 4–3.

Plant available potassium also has increased even though potassium fertilizers have not been applied except for sodium nitrate that contains 0.1% of this element. This notable increase in available potassium, that is not part of the organic compounds, remains active in the plant and is

Table 4–3. Nitrogen, phosphorus, and potassium in relation to organic matter content in soils of Chequén under three management systems.

Management system	Depth	Organic matter	Nitrate–nitrogen	Phosphorus	Potassium
	centimeter (inches)	%	———————— parts per million ————————		
Traditional	0–5 (0–2)	1.42	16	7	185
Wheat–oats	5–10 (2–4)	1.24	14	9	185
5 years	10–20 (4–8)	1.00	13	5	168
Pasture	0–5 (0–2)	4.56	20	32	237
+15 years	5–10 (2–4)	1.92	18	19	244
	10–20 (4–8)	1.14	21	15	255
No-till	0–5 (0–2)	5.32	64	51	325
Wheat–corn	5–10 (2–4)	2.84	58	46	280
7 years	10–20 (4–8)	2.24	27	5	232

easily released from the plant residue when the stubble is returned to the soil. I do not believe that the increase in potassium comes from the unavailable fraction because the minerals in the soil containing potassium are very resistant to weathering. The levels of available potassium to the plants has increased because of the residue that is left on the surface year after year. The rotation of crops including corn, lupins, wheat, and soybeans significantly helps to extract potassium from deeper parts of the profile; therefore, encouraging greater root development. I believe that this is the most reasonable way to explain this phenomenon.

In these analyses the greater content of organic matter occurred in the first 5 centimeters (2 inches) of the soil. It is interesting that the increase in organic matter content in the upper 5 centimeters (2 inches) was much higher after 15 years of pasture as compared with traditional tillage; however, it is even more interesting that a higher level of organic matter content was reached after only 7 years of no-till. In other words, the speed with which the soil managed with no-till is enriched with organic matter is practically double compared with that under pasture. Pasture is a soil management system long known for its efficiency in improving the fundamental soil resource.

It also was observed that plant availability of nitrogen, phosphorus, and potassium is greatly increased and appears to be directly related to organic matter, as shown in Table 4–3. For all of these reasons, I do not get tired of saying: **Management of crop residue, by keeping it on the soil surface, is very important for enriching the organic matter content in the underlying mineral profile. It not only promotes conservation, but also improves the productive capacity of the soil.**

With respect to cation-exchange capacity, analyses of Chequén soils indicate that there has been an important increase in their capacity to adsorb cations. In no-till with a corn–wheat rotation, the cation-exchange capacity increased from 11 to 18 milliequivalentes per 100 grams of soil, which is an increase of 63.6% compared with a plowed soil. The increase in base saturation is a phenomenon directly related to the increase in organic matter content of the soil more than a consequence of greater weathering of the mineral colloidal complex. Therefore, the cause of the increased CEC appears to result from the greater exchange capacity of the organic colloidal complex, as can be inferred from Tables 2–14, 2–15, 2–16, and 2–17.

4.4. IMPROVEMENT OF THE BIOLOGICAL CHARACTERISTICS OF THE SOIL

Improvement in soil biology that has been obtained with the introduction of no-till constitutes one of the most important achievements in the soils at Chequén. A few biological aspects

Active mesofauna such as soil slugs, earthworms, and sow beetles are visible just below the surface residue (1990).

Anthills on unplowed soil. The ants are efficient aerators, contributing to the improvement of the soil physical properties (1990).

that contribute significantly to the dynamics of soil fertility are described in the following paragraphs.

4.4.1. The Vesicular–Arbuscular Mycorrhizae and No-Till

By not plowing, but instead permitting a natural decomposition of residue on the soil surface, plus adequate fertilization and management, has allowed us to increase crop yields. It is easy to see that the fertilizer applied and the resulting yields are not in agreement with the high levels of phosphorus, potassium, and calcium established by the soil analyses made in recent years on Chequén.

After 7 years of planting no-till wheat in rotation with corn in severely eroded granitic soils, phosphorus is not applied and yet yields of more than 6 tonnes per hectare (89 bushels per acre) are obtained without reducing the available phosphorus levels in the soil. Corn yields were more than 12 tonnes per hectare (191 bushels per acre) with only 64 kilograms per hectare (57 pounds per acre) of P_2O_5 applied where the grain extracts more than 76 kilograms per hectare (68 pounds per acre). This indicates that in soils at Chequén, a significant reserve of nutrients is stimulated and used by the greater biological activity and organic matter.

Table 4–3 indicates levels of phosphorus under no-till are seven times those of a soil under a traditional system. Where and how do the plants extract phosphorus for their nutrition?

In the 1970s, I became interested in ectomycorrhiza because the literature indicated their importance in the nutrition of forest species. In effect, many of these higher plants are especially dependent on mycorrhizae in their physiological functions. Later, through review of the literature and direct contact with specialists, I learned that another group, the endomycorrhiza, also could be beneficial on the nutrition of lesser plants.

With the valuable cooperation of Drs. Fernando Borie and Rosa Rubio, professors at the Universidad de la Frontera, Temuco, we analyzed soils subjected to three treatments and identified different populations of vesicular–arbuscular mycorrhizae (VAM). Table 4–4 shows the results of the measurements made on samples corresponding to a depth of 20 centimeters (8 inches).

The great increase in VAM resulted from the generalized improvement of the soil microbiological activity and even a biological rearrangement, resulting from planting without plowing and by increasing levels of organic matter. The effect of no-till is notable in its contribution to the increase of VAM, both in the soil and on the roots of plants. We confirmed in the most recent measurements that 753 spores per 100 grams of soil were in the first 5 centimeters (2 inches) of the soil profile.

Humification of crop residue is the result of an integrated and successive action by fungi, actinomycetes, bacteria, and other microorganisms, plus the mesofauna that constitute the most

Table 4–4. Presence of vesicular–arbuscular mycorrhizae (VAM) in soils of Chequén under three management systems.

Management system	Depth cm (inches)	No. of spores per 100 grams of soil		Mycelium†		Root infection %	
		1986	1988	1986	1988	1986	1988
Traditional (plowed)	0–20 (0–8)	56	‡	3	‡	19	‡
Pasture	0–20 (0–8)	75	‡	2	‡	13	‡
No-till	0–20 (0–8)	216	460 (753)§	1	‡	27	‡

† 1 abundant; 2 common; 3 few.
‡ Not analyzed.
§ 753 spores in 100 grams of soil found in the top 5 centimeters of soil.

important mechanisms of degradation of dead tissues, and consequently the formation of humus (Zunino, 1983).

The fungi are the first microorganisms to participate in the destruction of vegetative residue. Its hyphae penetrate into the interior of the tissues and, with their enzymatic activity, achieve not only a mechanical, but also a structural rupture. Later, the associated bacterial population achieves a more complete decomposition producing the final degradation of the organic residue, part of which is metabolized to carbon dioxide and water. The compounds that are more difficult to decompose biologically, such as cellulose and lignin, can slow the action of the bacteria. These compounds serve as a substrate, however, so that the processes of humification continue as more residue is added. This was the first sign that we are practicing a more biological and natural agriculture. A more comprehensible response can be given after the analysis of mycorrhizae and other microorganisms.

The name mycorrhizae comes from the Greek, mukis means fungus, rhiza means root. The union of these organic bodies, the root and fungus, generates the development of spores associating itself with the root of the plants. As a result, a notable, completely natural symbiosis is initiated.

The VAM have been studied intensely by scientists for several decades. Their observations begin to have value for farmers when they practice a more rational and biological agriculture.

No-till and the VAM are mutually strengthened to give rise to an agriculture that is more conserving and productive. I believe that this symbiont is largely responsible for the greater levels of soluble phosphorus in the soil and that the rapid growth of this symbiont is the result of a system highly interdependent on other biological processes.

VAM are microscopic organisms that cannot be seen by the naked eye; therefore, a laboratory is needed to determine their existence and quantity. The endotrophic action is attributed to VAM's activity located within plant cells. After 10 days infestation of the host plant, the VAM generate their functioning parts, one of them is the arbuscular. This organ produces an endogenous transfer of nutrients captured by the hyphae, that are extra body projections on the plant, but that belong to the fungus. In this way the plants infected by mycorrhizae can obtain better nutrition because the hyphae can extend about 8 centimeters (3 inches) beyond the normal growth of the root hairs. Thus, they can absorb moisture and nutrients from the surrounding area and transport them to the interior of the plant. It has been determined that one centimeter (0.4 inches) of infected root has more than 1 meter (3.3 feet) of hyphae, which means a substantial increase in the capacity of the plant to absorb nutrients (Borie and Rubio, 1986).

The other body of the mycorrhizae, called the vesicule, permits its subsistence because these organs are small accumulators of lipids that are important in the nutrition of the fungus. The VAM are not capable of absorbing carbon from the air because they cannot capture sunlight and thus cannot photosynthesize. They also are not saprophytic fungi; that is, those that obtain their nutrition from organic bodies in a state of decomposition, nor are they pathogens that attack living organisms. Their energetic nutrition is obtained as a product of their symbiosis with the host plant. The plant feeds the fungus, and the fungus feeds the plant. This association is one of the great perfections of nature in the survival and ecology of living beings.

Experiments with radioactive phosphorus (P^{32}) in which the absorption of phosphates has been quantified for plants infected with mycorrhizae as well as for those not infected have shown that both obtain phosphorus from the soluble source. Scientists, add, however, that it is important to continue with this research to determine if the plants can absorb insoluble phosphorus by other means.

The high levels of available phosphorus (50 parts per million) of the soils at Chequén indicate that the VAM can act on insoluble or occluded phosphate. In this respect, Gianinazzi-Pearson

(1986) says that the plant and the VAM symbiont hyphae can generate enzymes or chemical compounds capable of making insoluble forms of phosphate available and thus transforming them for assimilation by plants.

Plants with roots infected with mycorrhizae can absorb four times more phosphorus than those without mycorrhizae, which indicates the high dependency of most plant species with this symbiont. Phosphorus, as the phosphate ion (PO_4^{-3}), is displaced towards the roots of the plant by diffusion. During this diffusion it can be captured or fixed by soil mineral or organic colloids that have high levels of calcium, aluminum, and iron (Azcon, et al., 1980); however, the movement of phosphorus in the soil is so slow that, in practice, only the phosphorus that surrounds the roots is available. Because of this, the assimilation of the necessary quantities of phosphorus for plant nutrition depends on the exploration capacity of the roots. Therefore, the greater capacity of the roots infected with mycorrhizae to absorb phosphorus results from a greater absorbent surface provided by the hyphae of the VAM that can explore a greater volume of soil (Borie and Rubio, 1986).

Farmers, knowing the benefits of the VAM, would like to have these and other microorganisms in the soil, but they do little or nothing to accomplish this. Unconsciously, they look for higher production by acting negatively on their environment. Without a doubt, the physical cultivation of the soil seriously diminishes the population of this fungi. Some chemically processed fertilizers whose availability and functionality are changed can become toxic or can seriously diminish the population of the VAM fungi. For this reason, the use of monocalcium phosphate fertilizers should be avoided on soils with pH 6.2 or less. Tricalcium phosphates should be used instead. It has been demonstrated that some fungi of mycorrhizae can use rock phosphate better than other sources and can optimize the growth of the plants in poor soils (Sieverding et al., 1984).

Other investigators, such as Juo and Kang (CORFO, 1986), express that in soils similar to those of Chequén (Alfisols) with a pH near 6.0, the VAM can improve use of phosphate rock by plants.

Mycorrhizae act as agents for improving the nutritional function of plants, especially where moisture or some chemical elements are deficient. Most plants achieve this symbiosis with the VAM; however, families, such as the Cruciferae (canola, radish), Chenopodiaceas (sugar beets), Carofilaceas, Cyperaceas, and Junglandaceas, do not have mycorrhizae (Borie and Rubio, 1986). In this case the VAM are introduced into the roots, but the plant quickly aborts it. Therefore, these crop plants are probably more dependent on fertilization in general, particularly on phosphate fertilizer.

The presence of VAM in legumes is of special importance. These atmospheric nitrogen-fixing plants benefit greatly from this symbiont. The legumes first become infected with mycorrhizae and later, strengthened by this primary symbiosis, produce nodulation by bacteria of the genus *Rhizobium*. With this mutualistic symbiosis the plant can achieve its complete development.

The VAM in plants also functions to better control pathogens. In wheat crops the attack by root fungi, such as *Fusarium* sp. and *Gaeumannomyces graminis*, is reduced if the wheat is infected with mycorrhizae (Sieverding et al., 1984).

Mycorrhizae aid soil fertility. If their environment is not mechanically disturbed by tillage, they strengthen and proliferate, benefiting other microorganisms. This improves the chemical, biological, and ecological aspects of the soils.

4.4.2. Nonsymbiotic Nitrogen Fixation and No-Till

As stated in Section 2.10.2.2., it is important and tremendously encouraging to point out the effect of no-till on the behavior of free-living microorganisms. Studies at the Center of Research in Biotechnology and Microbial Ecology, Buenos Aires, Argentina, have confirmed the positive effect of no-till.

Agronomist Alejandro Cariola studied these microorganisms in the soils at Chequén under three management systems. He sampled three sample depths using the Winogradsky technique of shaped soil plates as modified by Molina and Sauberán (Cariola, 1990). Using treatments with different forms of phosphorus as an activating agent with and without *Azotobacter* sp. inoculation, he observed the proliferation of colonies of this bacteria. Scientists attribute exceptional nitrogen-fixing qualities to this free-living microorganism when the environmental and nutritional conditions are suitable. They believe that it can be as important as that fixed by *Rhizobium* sp. on legumes by the symbiotic means.

Although the first phase of the study did not quantify the nitrogen supplied by the *Azotobacter,* this bacteria notably developed in soils under no-till. This was especially the case in the top two horizons analyzed, indicating the abundance of assimilable phosphorus in comparison with the soils under traditional management where the absence of assimilable phosphorus was measured. The abundance of assimilable phosphorus in no-till clearly agrees with the analyses of the soils at Chequén made every 2 years.

Other important considerations mentioned by Cariola (1990) confirm many ideas suggested in this book. One idea is that phosphate fertilizers normally used in traditional agriculture do not always yield expected results. It is logical then, that the form of phosphate best used by the *Azotobacter* in the study was disodium phosphate. Its buffer capacity neutralizes the acidity indicated by soil analysis. Also, the forms of organic phosphate are practically absorbed directly. The study also establishes that the soluble phosphates (triple superphosphate and ammonium phosphate) inhibit the action of this microorganism by generating acidity in the immediately adjacent area (pH near 1.9 adjacent to the granule).

Bradley, Sieling, and Dalton, mentioned by Cariola (1990), indicate that when the insoluble phosphorus is combined with iron and aluminum, the addition of organic material can free the phosphorus, making it available to plants.

The results of these observations confirm the necessity of managing the residue and fertilizer efficiently, thereby promoting soil microbiology since crop residue constitutes the basic food substrates for microorganism survival.

4.4.3. Effect of No-Till and Crop Residue on Soil Activity of Fungi and Bacteria

Dr. Fernando Borie has determined that the greater quantity of organic matter observed in soils under no-till and pasture has increased the biological activity, compared with soil under cultivation. As shown in Table 4–5, in the first 5 centimeters (2 inches) of soil, the phosphatase activity in the soil under no-till has increased 5 times, the fungi 2.3 times, and the bacteria 5.9 times compared with a traditional system. The comparative increase is reflected in greater fertility, especially in the high quantity of available phosphorus resulting from the increase in phosphatase activity. In the same table, fungi and bacteria that solubilize tricalcium phosphate have increased 4 and 16 times, respectively, under no-till compared with the traditional plow system.

The presence of fungi and bacteria that solubilize tricalcium phosphate is very important because of the quantity they can produce in soils that have a large capacity to fix water-soluble phosphates. These microorganisms require an environment appropriate for their proliferation. This observation agrees with that recommended by Sieverding related to using tricalcium phosphates in fixing soils. The resulting benefits that these microorganisms provide are significant (Sieverding et al., 1984).

Table 4–5. Comparative biological activity under three soil management systems on Chequén.

Management system	Depth cm (inches)	Phosphatase activity μg PNF per gram†	Fungi ($\times 10^5$) Total	Fungi ($\times 10^5$) Soluble‡	Bacteria ($\times 10^5$) Total	Bacteria ($\times 10^5$) Soluble‡
Traditional	0–5 (0–2)	133	1.6	0.3	1.0	0.3
	5–10 (2–4)	200				
	10–20 (4–8)	168				
Pasture	0–5 (0–2)	452	2.2	0.4	2.4	3.0
+15 years	5–10 (2–4)	280				
	10–20 (4–8)	178				
No-till	0–5 (0–2)	680	3.7	1.2	5.9	5.0
7 years	5–10 (2–4)	301				
	10–20 (4–8)	340				

† Fungi and bacteria that solubilize tricalcium phosphate.
‡ Micrograms of Paranitrophenol per gram of dry soil.

4.4.4. Effect of Earthworm Population

More than 2000 years ago in Egypt, the Pharaohs, by means of edict, protected earthworms, prohibiting their removal from the natural environment. Naturalist Charles Darwin studied these annelids in the past century concluding they were useful in the fertility and improvement of the soils physical characteristics.

In the processes of soil formation and improvement of its fertility, earthworms (*Allolobophora* sp.) have greatly influenced the management of no-till. They are a spontaneous response to ceasing tillage of the soil and are even more abundant when food resources, such as crop residue, are left on the surface.

Tools that disturb the soil also destroy the natural habitat of worms. Normally, the greatest population of earthworms is in the first 5 centimeters (2 inches) of the profile, which is most affected by tillage and constant passage of agricultural machinery.

The earthworm population was measured 7 years after initiating no-till on Chequén. The results are shown in Table 4–6. The population was 36 times greater in the no-till system compared with that in a soil under cultivation (Crovetto, 1989).

In counting worm populations, sow bugs (*Porcelia chilensis*) and other arthropods of the mesofauna also were observed. Again, this indicates the necessity of maintaining the stubble over the soil to achieve the benefits of increased biological activity.

With long-term no-till, earthworms are extremely important in humification of the existing residue on the soil. Residue left on the surface is digested by the worms when moisture and temperature conditions are adequate. This phenomenon, with a particular effect on the fertility of the soil, is an active cooperator in the breakdown of organic matter and in the work of the soil microorganisms. When the carbon/nitrogen ratio of crop residue is rectified by nitrogen, the biological decomposition is benefited because the initial processes are accelerated. In no-till, it is extremely important to transform residue into humus at a rate that does not allow the accumulation of organic materials for more than three years. The rate of decomposition depends mainly on: (1) type of organic residue, (2) surface moisture and temperature of the soil, and (3) biological activity of the soil. The process of humification is more rapid in humid and warm climates than in temperate climates. Therefore, the greatest quantity of residue should be left on the soils in these humid, warm climates and surface nitrogen added as needed.

Worms and other organisms of the soil mesofauna are activated by organic matter originating from residue (1991).

Worm casts produced on soils with abundant residue are a natural, low cost, efficient fertilizer (1991).

Table 4–6. Population of earthworms (*Allolobophora* sp.) under three soil management systems on Chequén.

Management system	Depth cm (inches)	Number of worms per square meter
Traditional	0–5 (0–2)	1
5 years wheat	5–10 (2–4)	2
	10–20 (4–8)	0
Pasture	0–5 (0–2)	28
+15 years	5–10 (2–4)	12
	10–20 (4–8)	1
No-till	0–5 (0–2)	67
+7 years	5–10 (2–4)	40
Wheat–corn	10–20 (4–8)	1

A good example of the effect of earthworms in a plowed versus a no-tilled operation established in a tropical Alfisol of western Nigeria is shown in Table 4–7. As indicated, the production of casts is four or five times greater in no-tilled soils compared with soils under tillage. As a result of the increase in the earthworms activity, the quantity of pores was greater and the bulk density was dramatically lower. The weight of worms excrement can be up to 50 tonnes per hectare (22 tons per acre). Casts are resistant to raindrop impact and contain a considerably greater quantity of organic matter than the surface soil (Lal, 1976).

One of the most favorable effects of earthworms in the process of decomposition of organic residue is the construction of galleries or labyrinths in their search of nutrients. Galleries are vitally important in the successful establishment of no-till, because they not only furnish space for plant root development, but also provide air, water, and nutrients in the form of humus. The microbiology of the soil is benefited and is strongly stimulated in the walls of the galleries. In this regard, some researchers indicate that large amounts of the phosphatase enzyme are in worm galleries as a result of the presence of humus generated by earthworms, as well as the greater biological activity that is generated in these sites.

4.5. NO-TILL AND THE TAKE-ALL DISEASE IN WHEAT SEEDINGS

In earlier chapters the necessity of increasing the microbial mass of the soil, not altering the structure of the soil, and allowing the decomposition of the crop residue on the soil surface was stressed. These few words probably summarize ones experience with a fertile soil and coincides with the philosophy of soil management encouraged by no-till. Just observing the pedogenetic phenomena shows that the soil is undergoing something unusual. The first indicator is a greater microbial population.

Table 4–7. Effect of crop rotation on worm activity (Lal, 1976).

Rotation	Number of worm casts, per square meter[†]		Weight of worm cast, t/ha	
	No-till	Conventional	No-till	Conventional
Corn–corn	1060	90	41.3	3.5
Corn–peas	1220	372	47.6	14.5
Soybeans–soybeans	42	3	1.6	0.1
Peas–peas	28	36	1.1	1.4

† To convert cast per meter square to cast per acre, multiply by 4032.26.

Table 4–8. Yields of wheat on wheat in relation to take–all (*Gaeumannomyces graminis*) fungal attack.

Seeding year	Traditional		No-till†	
	tonnes per hectare	(bushel per acre)	tonnes per hectare	(bushel per acre)
1982	2.6	(39)	2.5	(37)
1983	2.5	(37)	3.2	(48)
1984	3.0	(45)	3.8	(57)
1985	4.8	(72)	5.6	(83)
1986‡	3.7	(55)	4.1	(61)
1987	3.0	(45)	4.1	(61)
1988	2.8	(42)	3.8	(57)

† The wheat stubble did not exceed 3 tonnes per hectare (2700 pounds per acre) at seeding time.
‡ Delayed seeding due to wet fall.

In 1980, I read *Soil Dynamics* by the French soil chemist A. Demolon. In this book, microbial sociology is mentioned as a way of explaining the interrelationship of the different biological components of the soil. Demolon states that there are both the phenomena of association and antagonism in the soil, where independent, interrelated groups exist. In addition, there are other groups capable of generating auxins and/or toxins, thereby stimulating the development of other organisms or inhibiting or destroying pathogens for their own benefit (Demolon, 1965).

Tillage of the soil disturbs and diminishes this microbial and biological society; however, untilled soils tend to form the biological society again, achieving in some measure a similarity to those that exist in forests and pastures.

A large part of plant health is attributed to the relationship between the soil, plant, and natural equilibrium of the biological components of the soil. The alteration of this delicate equilibrium by misguided human actions can modify some components of the microbial population and cause the presence of pathogenic organisms in the soil.

In *Soil Dynamics*, Demolon explains that take-all disease (*Gaeumannomyces graminis*), an endemic root pathogen in the majority of the soils of the country, ceases its action when organic fertilizers are added. Soils that have been mistreated by mechanical cultivation, and thus have a lower level of organic matter and microbial population, are prone to attack by the pathogen, because these microorganisms that control the pathogen population have diminished (Demolon, 1965). In virgin soils or those recently tilled, this pathogen is latent and does not damage pasture grasses; however, plowing and disking disseminate the spores of the fungus, and at the same time, reduce the natural microbiological resistance of the soil.

On Chequén, the behavior of take-all disease in the soil with a monoculture of wheat was observed under both a traditional and no-till system. In the former, the infestation affected an average of 18% of the plants that had white, erect spikes, and emaciated grain causing premature death of the plant. The damage during the first few years of the study was 22%, diminishing toward the end of study to 14%. In the no-till trial of wheat on wheat damage was observed only 2% of the plants. This experience verified the thoughts of Demolon, and proves that the act of not plowing the soil permits the gradual recuperation of a natural biological equilibrium that controls the take-all pathogen.

It is interesting to analyze the yields shown in Table 4–8 comparing the behavior over time. Figure 20 shows the data as well as the difference between no-till and traditional tillage. Starting with the second year of no-till, an appreciable recuperation is produced, with a clear tendency toward the divergence of the curves. This indicates greater productivity in no-till.

It should be stated clearly that the practice of monoculture of wheat on Chequén has been done only to demonstrate experiments. Under no-till it is initially attractive to practice monoculture, however, it is not recommended as an agronomic system of plant production.

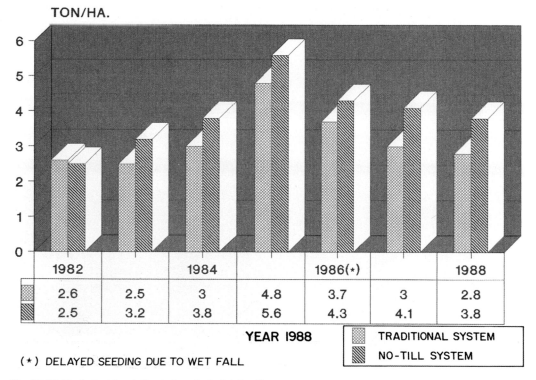

	1982		1984		1986(*)		1988
▦	2.6	2.5	3	4.8	3.7	3	2.8
▨	2.5	3.2	3.8	5.6	4.3	4.1	3.8

YEAR 1988

(*) DELAYED SEEDING DUE TO WET FALL

▦ TRADITIONAL SYSTEM

▨ NO-TILL SYSTEM

Fig. 20. Yield of wheat in relation to fungal attack take-all.

BIBLIOGRAPHY

Azcon, C., G. Aguilar, y J.M. Barea. 1980. Micorrizas. *Investigación y Ciencia* (España) 4:8–16.

Borie, F., y R. Rubio. 1986. Micorrizas. Su rol en ecologia y nutrición vegetal. *Próxima Década* (Chile). pp. 77–95.

Cariola, A. 1990. Influencia de la labranza cero en la fijación biológica de nitrógeno no simbiótico. *En* Primeras Jornadas Binacionales de Cero Labranza, Chequén, Florida (Chile). pp. 77–95.

CORFO (Chile). 1986. Evaluación agronómica de la roca fosfórica de Bahia Inglesa. Gerencia de Desarrollo, Comité de Sales Mixtas. 6 pp.

Crovetto, C. 1989. La cero labranza en siembras de trigo y su influencia en el medio edáfico en suelos erosionados de la Cordillera de la Costa de Chile central. *En* Seminario Técnicas de Riego y Conservación de Suelos para el Sur de Chile. INIA. *Serie Remehue* No. 9. pp. 157–170.

Demolon, A. 1965. *Principios de agronomia.* Tomo I. Crecimiento de los vegetales cultivados. Traducción de la 5ta. edición francesa por José Pérez Malla. Ediciones Omega S.A., Barcelona, España. pp. 419–436.

Gavande, S.A. 1973. *Fisica de Suelos. Principios y aplicaciones.* Editorial Limusa- Wiley, S.A., México. 351 pp.

Gianinazzi-Pearson, V. 1986. Les mycorrhizes: moyen d'ameliorer l'utilisation des engrais phosphates. *Fertilizants et Agriculture* (Francia) 92:3–10.

Lal, R. 1976. No-tillage farming in the tropics. *In* No-tillage Research Reports and Reviews. R.E. Phillips, G.W. Thomas, and R.L. Blevins, (Editors). University of Kentucky, College of Agriculture and Agricultural Experiment Station. Lexington. pp. 103–151.

Sieverding, E., M. Sanchez, y N. Bravo. 1984. Investigaciones sobre micorrizas en Colombia. *En* Memorias del Primer Curso Nacional sobre Micorrizas. Universidad Nacional de Colombia. Palmira, Colombia. pp. 15–38; 64–67.

Tisdale, S.L., W.L. Nelson, and J.D. Beaton. 1985. *Soil fertility and fertilizer.* Macmillan and Co., Ed., 4th Edition. New York. pp 662 667.

Zunino, H. 1983. Suelos ecuatoriales. Ecologia microbiana, acumulación de humus y fertilidad en suelos alofánicos. Sociedad Colombiana de la Ciencia del Suelo. XII(1):23–28.

CHAPTER 5

Productivity and Profitability

If you have little soil,
Make it grow.
If you have much,
Grow yourself to deserve it

ALBERTO GALLINAL

The no-till system permits harvesting and planting at the same time, promotes better resource usage, and therefore increases production.

It is not easy to talk about cost and profitability in crop production. Showing results of a specific situation is difficult because of the many variables. Furthermore, mistakes can be made that are discouraging, and the soil can be affected positively or negatively for the intermediate or short-term. The different systems that are implemented in crop production and the different soil types in the same field can lead to mistakes. The costs and profitability can be different in two neighboring fields with the same soils and planting time, indicating that the agronomic systems implemented are different.

Efficient input application for a specific crop is very important. The variation of inputs and application methods can result in significant differences in the final farm productivity and profitability. Many times there are good fields without a corresponding field to compare its productivity. The opposite also can happen. Chequén has shown, under certain conditions with no-till, that **low fertility soils can be more profitable than fertile or irrigated soils.**

Although the best opportunity for success in farming is attributed to good soil quality, it can be more profitable if other aspects, such as land cost and interest over capital invested, are considered for a poor quality soil. If adequate technology is applied to the productive system of an inferior quality soil, it is possible that its profitability will be higher than for a better quality soil. What has basically been described in this book is that a deteriorated soil can be productive in the short-term. It also has been emphasized that the soil's fertility is not only due to the quality of the minerals that make up the clays, but also to good conservation and agronomic management.

One aspect that is even more important than the quality of the soil is the climate, which can only be partially managed. It can be more productive to have a soil with inferior quality under adequate climatic conditions than a good quality soil under inadequate conditions. That is why summer irrigation in soils under dry and hot climatic conditions is very important. Therefore, the corn productivity at Chequén is possible because of sprinkler irrigation. Without it, the corn cannot be planted because the surface soil is generally dry after October. The fall crops, in years of normal precipitation, have high yields and are profitable. In very dry years, no-till has shown that crops planted in that season (May to June) also are profitable. Good use of scarce rains is very important. Rainwater is necessary for the seeded land to begin development. The residue that helps to hold soil humidity also is very important.

Another economic variable is the commuting distance to markets, which can have a big influence if the roads can be used in all seasons. Interior roads and electrification of the farmstead can reduce the production costs and make farming more efficient and enjoyable. Since the beginning of Chequén, we understood that it was fundamental to have an adequate infrastructure, which materialized as production increased. Adequate availability of equipment in the field can mean a timely and efficient operation. Chequén is not in a farming area. Equipment that can be leased comes from more than 100 kilometers (62 miles) away. Because of this disadvantage, we must own equipment to be sure it is available at the time it is needed. This required more working capital; however, the higher costs of equipment investment are compensated quickly with more efficient and timely operation.

At Chequén, reinvesting profits to gradually achieve adequate interior roads, electrification, drinking water, communication media, equipment, machinery, buildings, sheds, dwellings, reservoirs, irrigation equipment, and other investments, has been very important.

A capital that cannot be easily seen is the increased yield and reduced production cost. The increase in organic matter levels in soils of Chequén constitutes an investment that is increasing every year with lower operational cost. This dramatic high level of organic matter is the reason for constant yield increases during the last 16 years. Enough has already been said in this book

Conservation and productive use of the soil resource should be the goal of anyone that has a direct relationship with the
  land. Steep, sloping land should not be an obstacle to achieve this goal.

about the role of increased organic matter in the soils under no-till and its vital role in the present
and future farming economy.

The physical behavior of the clays at Chequén just below the new surface soil was not discussed fully. I estimate that the colloidal illuviation process, for both minerals and organics, has a
strong pedogenetic effect on the old eroded soil. With this I want to clearly state that the no-till
benefits of leaving the stubble on the soil are not only in the surface layer, but also in the lower,
degraded horizons. As a result, if this production system continues, the physical conditions will
continue to improve with additional economic benefits resulting in the future.

The more relevant phenomenon that happened to the soils at Chequén is the notable
improvement of the erodible soils edaphic characteristics. With the use of the no-till system, soils
in Capability Class VI and even VII sustain yields similar to the soils in Classes II or III. The soils
of Classes VI, VII, and VIII are classified as nontillable according to this classification principle
and have low productivity mainly because of steepness. The steep soils that are tilled have a serious risk of erosion that significantly reduces their natural capacities. In an opposite way, no-till
presents the possibility to plant very steep soils now classified as untillable and yet achieve high
yields, while controlling soil erosion. All of this shows that the original Land Capability
Classification System has little, if any, relation to a properly managed no-till system; however, the
Land Capability Classification System was devised prior to the development of no-till technology. The productive economic potential is superior and is clearly supported by a more than adequate protection of the soil. This fact has a great economic importance when it is compared with
soils with low productivity that are tilled. **The great economic advantage of no-till is that low
value soils can be brought into higher production without erosion risks, formerly the privilege of only high value soils.**

5.2. The Value of What is Lost

Soil loss can be estimated by the use of soil erosion prediction tools. For soils with charac teristics and topography conditions like those at Chequén, a soil erosion rate would calculate to about 18 tonnes per hectare (8 tons per acre) per year when they are annually tilled. This number by itself is not significant, but if we try to clarify its meaning, we can conclude that the 18 tonnes per hectare per year of soil loss is equivalent to scraping off 1.4 millimeters (one twentieth of an inch) of top soil per year from the field surface. In terms of cultivable land this is a loss of a soil surface area [0.20 meters (8 inches) deep] equivalent to 69 square meters for each hectare (300 square feet per acre) of the field being eroded each year. When these numbers are extended to other regions, the losses can be very high. For example, in a 100 hectare (247 acres) field with soils in the process of 18 tonnes per hectare degradation, the losses can reach more than 0.70 hectares (1.7 acres) from the field per year. In about 4 years, the 100 hectares will be reduced to 97.2 hectares (240 acres) of its productive capacity. To this fact we have to add that the soil lost is the surface soil. Seeds and fertilizers also are lost. These fundamental agronomic goods convert directly into an important basic economic value that is no longer available.

Certainly the productivity at Chequén, considering its geoecological situation, is based on the fact that no-till leaves the stubble on the soil. Without doubt, this has been the cultural–physiological basis for the economic success of Chequén. We could not show the mentioned results today if we had not considered the crop stubble in addition to no-till. The use of stubble avoids harsh farming practices that involves traditional soil mining; while no-till provides the needed conservation.

The Argentinian Friends of the Soil Association (1987) records and publishes the loss of energy from stubble and residue burning in relation to farm exploitation. If it is assumed that straw contains 4200 calories per kilogram (1900 calories per pound) of energy, then burning of 4 tonnes per hectare (3572 pounds per acre) wheat stubble per year can represent a loss of 16,800,000 calories per hectare (6,786,800 calories per acre) per year. This is equivalent to the senseless waste of energy produced by 1680 liters (444 gallons) of petroleum.

5.3. The Value of What is Gained

Returning to our initial setting regarding the new superficial horizon achieved with no-till management, the 1 millimeter of soil organic material that develops annually not only means eliminating soil losses by erosion, but also the gaining of an equivalent of 50 square meters per hectare (217 square feet per acre) per year of cultivable soil 20 centimeters (8 inches) deep. The same analysis criteria are used to estimate the gain as was used for soil loss erosion. If we take this idea to the same 100 hectares (247 acres), a new surface area 20 centimeters deep of 0.5 hectares (1.25 acres) per year is formed and in 4 years the 100 hectares (247 acres) will be converted into 102 hectares (252 acres)

We must clarify that this situation does not happen in reality because physically the 100 hectares that are considered in the example cannot grow or decrease on the same site. What should be very clear is that in economical terms the 100 hectares can now produce the equivalent of 102 hectares of the original soil.

The gain of soil that occurs with no-till is in the same proportion as the loss of soil by erosion without no-till.

If we can place a value on a beautiful and rich landscape, I believe it would be priceless. This is where you see life from different angles. Each year different organisms emerge

On windy days tillage can result in serious soil losses by wind erosion. We also can see the difficult environmental conditions under which the operator and equipment works (1990).

High flow of the Mapocho River in Santiago. The improper use of natural resources not only damages the farmers, but also the whole community (1982).

with their natural surrounding. These organisms perform a specific, complex, and complete function for the environment.

5.4. OUR RESOURCES AND THEIR USE

The integral use of resources that are generated from farming have been extremely important in Chequén's development.

The wood industry residue, such as sawdust and wood shavings, is important in the poultry production, cropping, and cattle enterprises. For example, wood shavings are used as bedding in the poultry barns, while a mixture of sawdust and wood shavings is very useful for the improvement of soils in gullies. During the winter, sawdust generates heat for dwellings.

The by-products generated by the poultry enterprises are important in the production of grasses and forages. The poultry waste or chicken manure from caged layers is a fertilizer for grassland, increasing the beef production at a very economical cost. A similar situation happens with the poultry litter, which is used directly in feeding beef cattle.

A small hog enterprise also uses the residue from the poultry farm. The residue collected from a feed mill floor and chicken houses is used to feed hogs. This also reduces production costs.

The pruning and clearing of woodland contributes to a good source of winter food for Hereford cattle. These cattle consume the pine needles when there in not enough forage, but only if the tree trimmings are less than 2 months old.

Wheat and corn stubble also constitutes a good source of food for the beef cattle *under very specific conditions*, the residue can be reduced through direct grazing by cattle. There also is a slight gain of live weight. The corncob also can be used as cattle feed if it is finely ground and mixed with other by-products during the winter. This concentrated feed incorporates the chaff resulting from cleaning the harvested grain.

No-till provides a farmer the opportunity to increase cattle numbers and produce more beef per area. The higher quantities of residue produced in the system, if well used, help to reduce cattle feeding cost.

Because tillage tools are not required, the no-till system allows the farmer plant on the same day of the grain or silage harvest. This is economically important because it lets the farmer produce more in less time on the same unit of land. It has been proven at Chequén that efficient use of any residue, such as lupin and oat hulls, or residue from the grain cleaning machine or from the animal enterprises can mean an appreciable increase in the economy of farm production.

Herman Warsaw, a corn farmer from Saybrook, Illinois, achieved a corn yield of 23,210 kilograms per hectare (369 bushels per acre) in 1985. This is recognized as a world record production for corn without irrigation. Besides efficient conservation management, Warsaw adds 22,400 to 44,800 kilograms per hectare (10 to 20 tons per acre) of cattle manure produced on his farm to his soil. These high yields can to be highly profitable and also benefit the soils both now and in the future (Nelson et al., 1985).

The cost and profitability computation has not been necessarily a routine procedure for Chequén crops; however, at the end of the crop season, we at least determine the production costs and profitability of the main grain crops.

5.5. WHEAT AND CORN PRODUCTION COSTS AND PROFITABILITY

From 1988 to 1989 the corn production cost with sprinkler irrigation on a 30 hectare (74 acre) field with soils in Capability Classes IV, VI, and VII was equivalent to 5315 kilograms per

Deforestation, and later soil tillage, of Chequén soils to produce wheat increased deterioration of the soil until reaching the situation shown in the photo in 1955.

After 35 years of soil conservation with no-till, a crop production system was achieved on the same soil as that in the picture above. The corn variety Pioneer 3540 in this picture had a yield of 17,179 kilograms per hectare (273 bushels per acre), registered by agronomists Jorge Barahona and Pablo Pedreros of Pioneer Chile (1990).

The lumber by-products generated in Chequén are used in soil conservation. Sawdust is being applied to protect the soil from erosion in areas where gullies have been filled and smoothed (1985).

hectare (84.5 bushels per acre). Because the average production was 11,900 kilograms per hectare (189.2 bushels per acre), the profit was estimated on 6585 kilograms per hectare (104.7 bushels per acre).

Winter wheat planted in oat stubble during the same period had a production cost of 1920 kilograms per hectare (28.6 bushels per acre). With a yield of 3800 kilograms per hectare (56.6 bushels per acre), the profit was 1880 kilograms per hectare (28.0 bushels per acre).

The year 1990 was one of the driest in the 20th century and the driest in the last 20 years. That years wheat crop in corn stubble on a 20 hectare (50 acre) field had a production cost of 1840 kilograms per hectare (27.4 bushels per acre) and a yield of 5830 kilograms per hectare (86.9 bushels per acre), with a profit of 3990 kilograms per hectare (59.5 bushels per acre). No preplant herbicides were applied because no green weeds were in the corn stubble. The higher yield probably resulted from an early planting, good rainfalls in the fall, more efficient weed control, and fertilization based on calcium magnesium ammonium nitrate. This fertilizer is highly efficient in no-till, therefore the quantity applied per area is reduced, which results in economic benefit.

Other information on the benefits of no-till was provided by Roberto Velasco, agronomist from the Farming Research Institute. He researched the performance of an oats–wheat rotation at the Santa Rosa experimental farm in Pinto County, Ñuble Province. When the traditional planting system was compared with no-till, it was established that the number of operations needed before planting to achieve adequate crop establishment differ. Table 5–1 shows the differences.

Based on equipment time required per area, it was observed that no-till has a lower energy consumption, the operations are more timely and quicker, and agroecology benefits are greater. This substantiates the observations made in all previous chapters.

The cleaning and sorting of grains produce valuable by-products for livestock feeding and add value to the products (1990).

Sawdust and shavings are an excellent litter in poultry hatchery production. It is ultimately used for livestock feeding (1991).

Manure generated in poultry and hog enterprises is handled mechanically. This facilitates and accelerates the collection and transportation of the waste (1990).

The application of organic wastes in pastures is performed using mechanical spreaders (1990).

Corn silage is a good alternative for livestock feed during the season of low production of green grass (1991).

Livestock winter feeding is provided by grains and other farm by-products (1990).

Reinvestment in farming is fundamental for continuous growth (1989).

The wheat yields observed in this research were 4235 kilograms per hectare (63 bushels per acre) on the traditional system and 5110 kilograms per hectare (76 bushels per acre) in the no-till system; a positive difference of 875 kilograms per hectare (13 bushels per acre) for the no-till system.

Comparing the soil preparation and planting costs in the same research for the traditional and no-till systems, Velasco determined the values shown in Table 5–2, expressed in United States of North America dollars per hectare (Velasco, 1990).

According to Table 5–2, a higher cost is calculated for the no-till system of $18.87 per hectare ($7.61 per acre), To offset this additional cost, a yield increase of 130 kilograms per hectare (2 bushels per acre).was needed. The yields obtained in this research easily absorb this higher cost. Therefore, considering that the no-till yield was 875 kilograms per hectare (13 bushels per acre) more than the traditional system, the additional cost of 130 kilograms per hectare (2

Table 5–1. Comparison of the number of operations prior to planting wheat, between two systems: the traditional planting system and the no-till system; 1987 to 1990 period, Pinto County, Ñuble Province, Region VIII. Post planting management was the same (R. Velasco, 1990).

Operation	Tillage systems			
	Traditional	Hours per hectare	No-till	Hours per hectare
Stubble cutting	--	--	1	0.6
Cleaning and burning the stubble	1	--	--	--
Residue incorporation	2	2	--	--
Vibrating cultivator	1	0.6	--	--
Herbicide application	--	--	1	0.2
Total	4	2.6	2	0.8

Grain production is compatible with life, and only people can make it an enemy (1990).

Table 5–2. Comparison of the soil preparation and planting costs for wheat between the traditional and no-till systems; 1987 to 1989 period; Pinto Estate, Ñuble Province, Region VIII. The post planting management was the same (R. Velasco, 1990).

Month	Operation	Tillage Systems	
		Traditional	No-Till
		USA $ per hectare	
February	Cleaning and burning stubble	0.35	0.00
February	Stubble mowing	0.00	7.71
February	Offset disking	16.13	0.00
April	Chemical fallow†	0.00	42.94
May	Offset disking	16.13	0.00
May	Vibrating cultivator	9.68	0.00
May	Planting‡	17.66	28.15
Total		59.95	78.83

† Labor cost plus herbicide.
‡ Only labor costs.

bushels per acre) is deducted for a higher net profit of 747 kilograms per hectare (11 bushels per acre) for the no-till system compared with the traditional system. In the same research, Velasco points out that in the traditional system the cost was 2733 kilograms per hectare (41 bushels per acre) of the total yield of 4235 kilograms per hectare (63 bushels per acre), which means a profit of 1502 kilograms per hectare (22 bushels per acre). In the no-till system, the cost was 2863 kilograms per hectare (43 bushels per acre) of the total for a yield of 5110 kilograms per hectare (76 bushels per acre), representing a profit of 2247 kilograms per hectare (33 bushels per acre).

Velasco concluded that the no-till system has a comparative economic advantage, which when added to the most important ones, such as soil conservation and energy savings (petroleum), strongly implies that this is a practice that farming should chose for its own if the productivity of the soil is to be saved for following generations (R. Velasco, 1990, personal note).

The winter wheat crops in Chequén are not irrigated, depending on the seasonal rainfall, especially during fall and spring. The yield of wheat in rotation with lupins and corn during a year with normal precipitation has reached 7200 kilograms per hectare (107 bushels per acre).

Under a corn monoculture in soils with low moisture content, planting winter oats in the corn stubble is a normal management practice. The corn stubble is not used for cattle feeding, but for soil fertility improvement. Oats can produce 3 tonnes per hectare (2700 pounds per acre) of dry matter between May and September and are only occasionally used for cattle feed.

The positive effects of no-till cannot be measured in a short time. For this reason some agronomic and economic results are difficult to see after only 1 to 3 years of no-till. The no-till system is based on a natural process in which the minimum time to obtain objective results is the time needed to decompose the first stubble on the soil. To evaluate this process without considering these aspects, can prejudice the results of those supporting and promoting the no-till system.

Without doubt, the permanent use of stubble and crop rotation in a no-till system is fundamental to conserving and managing the soil in a productive way. I believe that the farming production at Chequén is based on conservation of the natural resources and is seriously supported by a complex economic system. It is important to point out that nothing could be achieved without the personal belief and commitment to achieve the proposed objectives.

I have the support of Chequén's laborers in my extensive work as a conservationist. Their efforts and dedication made possible the changes that we show with pride today. This can influence the future decision of the farmers from other places who visit us. The young people who participated in these changes are some of the most important achievements we accomplished. These

Chequén's farm laborers are the great performers in our conservation and production progress. They are the indivisible links of their own well-being (1991).

students, sons and daughters of Chequén's workers, get the highest marks in their respective schools. Most of them have completed their high school education, and some are continuing with professional careers. These valuable young people have a big future and are without doubt the most important harvest of Chequén Farm. I believe that some of them will continue in direct contact with the land, strengthening the work performed and teaching others a better way of life.

BIBLIOGRAPHY

Asociación Amigos Del Suelo. 1987. Resumen de 30 años de trabajo. Memoria 1957–1987. Buenos Aires, Argentina. 48 pp.

Nelson, W.L., e H.F. Reetz, Jr. 1986. Herman Warsaw e sua bem sucedida busca de maiores produçoes de milho. Do Journal de Educa çao Agronômica, Sociedade Americana de Agronomía. 15:2–8.

Aerial photo of Chequén Farm.

Traditional agriculture with animal-drawn equipment.

Productive wheat fields at Chequén Farm.

EPILOGUE

The patient and painful observation of the erosion from which my country suffers has brought me to write this poem. It is none other than the cry of profound suffering of our soil which, while being destroyed, begs mercy from its executioner.

Golden wheat you harbor death

Wheat, as you nourish my life
and murder my land
sweating and defenseless you leave me,
you get to my core
like mother's milk
and having you I am satisfied.

Golden wheat, bread of hearts
bitterness for those who give you life,
downhill l look at you,
your roots hanging,
my land split open
cries out its hopeless suffering.

Your grain heads in the wind
show the richness of life
and the pain of death
because you tear open my land to bear
you.

I love you true son of the earth,
prodigal and spiteful like no other
help your wounded lover
over the green darkness.

You fed the gold,
cut down hopes and killed
brother of greed, already I
cannot go on,
you are part of me,
let me live.

Tears from the sky,
sparkling emeralds
filling your body
killing my hope.
Beg of your undaunted suitor that your
slender body
does not fall buried
in an ashen shroud.

Reflect, then, l implore
on the evil you have done
leave your plowed bed,
cover my hills, l ask
without embracing death.

l long for you as a loyal friend
that your fleeting wealth
may burst from your soul
to relish you to eternity.

EPILOGUE

GLOSSARY

Abiotic: Nonliving, basic elements and compounds of the environment.

Absorption: The penetration of a substance into or through another, such as the movement of a gas in a liquid.

Actinomycetes: Organisms present in the soil that form a group of mold-like microorganisms that are generally considered intermediates between bacteria and fungus. They are significant organisms in the decomposition and stabilization of organic material. They adapt better to soil moisture and temperature stress than fungus and bacteria.

Adsorption: Adhesion of a substance to a solid or liquid surface.

Adjuvant: An inert material that is mixed with the active ingredients in a pesticide to improve the effectiveness of the mixture.

Aerobic: Growing in the presence of and absorbing oxygen to facilitate respiration.

Aggregation: The cementing or binding together of soil particles to form secondary units of aggregates or granules. Water stable aggregates are difficult to break and are of special interest to soil structure.

Allophane: Clay particles formed from volcanic ash rich in aluminum and silicate. These clays adsorb phosphorus, especially monocalcium phosphates.

Amendment: Compounds used in agriculture that are mixed with the soil to correct and improve the chemical, physical, and biological conditions. Two amendments most commonly used are lime (calcium carbonate) and gypsum (calcium sulfate).

Amino acids: Building blocks for the construction of proteins that contain amino groups ($-NH_2$). These groups are either basic like amines or acidic like carboxyl.

Ammonification: The biological process whereby nitrogen is converted from organic nitrogen material to ammonical nitrogen. The amino groups consisting of $-NH_2$ compounds change to NH_3 by the action of enzymes produced by microorganisms, more so in aerobic than anaerobic conditions.

Amphoteric: Chemical compounds that have the ability to neutralize both acids and bases. Also, can attract and bind either cations or anions.

Anaerobic: Growing in the absence of oxygen. Certain microorganism that live in the absence of molecular oxygen yet can still respire.

Anion: Ion with a negative electrical charge which during electrolysis is attracted to the anode, the positive pole.

Anthropogenic: Pertaining to the activity of humans or resulting from the actions of humans.

Arthropods: The group of invertebrate animals that have jointed bodies and limbs. Some members are insects and crustaceans.

Auxins: Organic substances that activate the growth of plants.

Bases: In chemistry, a group of compounds that have the property to form salts when combined with acids.

Buffer: Ability of the soil to resist abrupt changes in acidity or alkalinity in the soil solution. This is possible because of the presence of particular colloids, minerals, and organic compounds.

Calcination: Heating mineral and organic compounds to a high temperature that drives off volatile materials such as liquid or gas.

Carbohydrates: Hydrated carbon. Organic compounds formed by combining carbon, hydrogen, and oxygen. Sugars, starches, and cellulose are all carbohydrates produced by plants.

Cation: Ion with a positive charge that during electrolysis is attracted to the cathode, the negative pole.

Cellulase: An enzyme that hydrolyzes cellulose, which starts the decomposition of the cell. These are enzymes produced by microorganisms capable of destroying the tissue of plant cells and initiate the process of plant decomposition.

Colloid: Colloids are inorganic and organic matter having very small size and correspondingly large surface area per unit mass. Particles size (0.0001 to 0.0000001 millimeter) that in liquid appear to be dissolved, but are not, only held in suspension. These particles have a high capacity to adsorb ions.

Colloidal activity: Colloids have a high capability to absorb electrons and take on electrical charges, positive or negative, that provide the colloid with movement because of the attractive or repulsive forces.

Compaction: The act of uniting firmly. The process of becoming compact. In soils that are compact the density is greater with less pore space. It is a reduction of the pore space in the natural structure by the effect of mechanical action on the soil.

Cultivar: Plant species that are used as agronomic crops.

Cultivation: The labor practices necessary to control weeds, seed and fertilize the land, and make the plants grow and produce fruit.

Deforestation: Processes that lead to removal and destruction of the forest.

Denitrification: The chemical or biochemical reduction of nitrate or nitrite to gaseous nitrogen, either as molecular (N_2) or as an oxide of nitrogen (N_xO). In the soil this takes place when microorganisms use the oxygen of the nitrate compound for its metabolism.

Desertification: Processes that are conducive to creating a desert or conditions that exist in a desert.

Dicotyledons: Plants that have two cotyledons in the seed. These two cotyledons develop into two leaves at germination.

Ecosystem: The interacting system of biologic community and its nonliving environment. The interaction between the components of the environment produce an interchange of materials within the system.

Edaphatic: Pertains to soils and the relationship with plants and other organisms in the soil. Combined processes that pertains to the development of soils.

Electron: Basic elemental particle that forms part of the atom and constitutes a negative electrical charge.

Endemic plants: Plants that are restricted to a relatively small geographic area or to an unusual or rare type of habitat.

Endotrophic: Acquiring nutrient substances from within the cell. Mycorrhizae mycelium penetrates the host plant root cells and extends into the surrounding soil.

Enzyme: Protein substances that are produced in living cells which catalyses biochemical metabolic processes.

Evapotranspiration: The combined loss of water from a given area by evaporation from the surface of soil and plants and the transpired water loss from the plant.

Field capacity: The percentage of water remaining in the soil after it has been saturated and allowed to drain for 48 hours. The percentage of water loss upon drying in an oven at 105°C related to the weight of the soil after drying.

Gully: A channel cut by concentrated water runoff which water flows only during heavy rain or snow melt events. A gully is sufficiently deep that it would not be obliterated by normal tillage operations.

Humus: The more or less stable fraction of the soil organic matter remaining after the major portion of the animal and plant residues have decomposed. The organic materials are the results of a diametrical process that involves both death and life, building and breaking down.

Humification: The process whereby carbon of organic residue is transformed and converted to humic substances through biological and chemical processes.

Infiltration: The downward movement of water from the soil surface into the soil matrix.

Intrusion: Process whereby mineral mass or substance penetrates into another mineral and produces a physical or chemical change in the minerals.

Ion: An atom that has become electrically charged by the loss or gain of one or more electrons.

Latency: A state or condition that exist of being present or capable but is not now manifested or active. An example is a weed seed that is present, but does not germinate because current conditions, yet can when conditions become more favorable.

Leaching: The process of washing soluble compounds through the soil with percolation water. The soluble material can be deposited in a lower part of the soil profile or move into the ground water.

Lignin: Noncarbohydrate, organic material that constitutes, along with cellulose the structure of plant cell walls.

Lysimeter: A device to measure the quantity and quality of water movement through the profile of the soil.

Meristem: Specialized plant tissue at the tip of roots, stems, and branches where cells are rapidly reproduced and differentiate into other specialized cells.

Mesofauna: Soil animals that are more developed than microorganisms, such as nematodes, small insect larvae, and microarthropods.

Metabolism: Transformations of material that occurs in cells of living organisms whereby carbon and protein material are changes to produce energy. Metabolism occurs in two different phases: the anabolism, or building of living substances and catabolism, the destruction of living substances.

Mineralization: Transformation of a compound from an organic form to an inorganic form as a result of microbial metabolism.

Mycelium: A protrusion from the body of a fungus that can penetrate soil and organic material to extract nutrient from the host.

Nitrification: Transformation of ammonium to nitrite and nitrate in aerobic conditions by nitrosomonas and nitrobacter bacteria.

Osmosis: Tendency of a fluid to pass through a semipermeable membrane, like the wall of a living cell, into solutions of higher concentrations so as to equalize the concentration on both sides of the membrane.

Oxidation: Chemical reaction by which the quantity of oxygen of a body or compound increases and the number of electrons of an atom is decreased.

Panicle: An open inflorescence of the grass family.

Pathogen: An organism that causes disease. An example would be plant pathogens that cause plant disease.

Pedogenesis: A process that involves formation of the soil in relation to the environment through time.

Percolation: Downward movement of water through the soil matrix.

Permanent wilting point: The water content of the soil, on an oven dried basis (105°C), at which plants wilt and fail to recover after rewetting. The point at which plants can no longer absorb moisture from the soil.

Phenology: The study of the relations between climate and biological phenomena. The flowering of plants or activity of insects are part of the phenologic studies.

Photosynthesis: The synthesis of carbohydrates from carbon dioxide and water by chlorophyll using light as energy with oxygen as a by-product.

Physiological maturity: Fruits that have reached a state of maximum dry weight accumulation and developed a condition that would produce germination of the seed.

Physiology: The science that studies the processes and functions of living things.

Phytotoxic: Any chemical agent injurious to plants.

Protein: A group of colloidal compounds of high molecular weight that contain carbon, hydrogen, oxygen, nitrogen, and sulfur. These compounds are basic to the composition of living things, constituting the substances of enzymes, cells, hormones, and other substances important to life.

Proton: Positively charged particle that forms part of the nucleus of the atom. Its charge is equal in absolute value to that of the electron, but of the opposite sign (positive). The number of protons of each element is always the same as the atomic number for that element.

Reduction: Chemical reaction opposite of oxidation. The quantity of oxygen of the body or compound is decreased and electrons are gained by the compound.

Rhizosphere: The zone of soil and air immediately adjacent to plant roots.

Saturation (chemical): Solution that is in equilibrium with the solute at a specific temperature. When two or more compounds are combined in the maximum concentration possible and still remain in solution.

Seedling: Small plants following germination, but before flowers or fruiting occurs.

Soil physiology: Aggregate of biological, chemical, and physical processes occurring in the soil.

Soil structure: The combination or arrangement of primary soil particles into secondary units or peds. The structure can refer to the natural position of particles of the soil when it is in place without disturbance. The structure is an important characteristic that determines the potential for soil erosion.

Stomata: Leaf structure formed by two cells that can produce an opening to allow the interchange of gases and water vapor between the plant and the air.

Structural water: Soil water that chemically combines with the water of hydration of soil particles. This water is not available to plants and can only be removed by high temperatures. Also known as cohesion water.

Symbiont: Organisms living together in mutual association.

Symbiosis: Association of two or more dissimilar living organisms that are benefited mutually.

Systemics (herbicides): Herbicides that translocate to the interior of plants and distribute throughout the circulation system of the plant.

Toxins: Substances produced by living organisms, principally microorganisms, that act as poisons.

Traditional tillage: Use of the moldboard plow and other tillage equipment that reduce surface residue by incorporation and greatly alter the structure of the soil.

Transpiration: Function by which the aerial parts of plants release water vapor to the atmosphere.

Volatilization: Transformation of a solid or liquid compound to a vapor or gas. An example is loss of gaseous components such as ammonia nitrogen from manure and fertilizer.

Water table: Uppermost surface of the groundwater or the level below which the soil is saturated with water.

THE ELEVENTH COMMANDMENT

Walter C. Lowdermilk

Thou shalt inherit the holy earth as a faithful steward,
conserving its resources and productivity
from generation to generation.
Thou shalt safeguard thy fields from soil erosion,
thy living waters from drying up,
thy forests from desolation, and protect
thy hills from overgrazing by the herds,
that thy descendants may have abundance forever.
If any shall fail in this stewardship of the land
thy fruitful fields shall become sterile
stony ground and wasting gullies,
and thy descendants shall decrease
and live in poverty
or perish from off the face of the earth.